Great Geological Controversies

Great Geological Controversies

SECOND EDITION

A. HALLAM

Lapworth Professor of Geology
University of Birmingham

OXFORD UNIVERSITY PRESS

Oxford University Press, Walton Street, Oxford OX2 6DP

Oxford   New York   Toronto
Delhi   Bombay   Calcutta   Madras   Karachi
Kuala Lumpur   Singapore   Hong Kong   Tokyo
Nairobi   Dar es Salaam   Cape Town
Melbourne   Auckland   Madrid

and associated companies in
Berlin   Ibadan

Oxford is a trade mark of Oxford University Press

Published in the United States
by Oxford University Press Inc., New York

First published 1983
Second edition 1989
Reprinted 1992

British Library Cataloguing in Publication Data
Hallam, A. (Anthony)
Great geological controversies.—2nd ed
I. Geology, history
I. Title
551'.09
ISBN 0–19–858218–8
ISBN 0–19–858219–6 (pbk.)

Library of Congress Cataloging in Publication Data
Hallam, A. (Anthony), 1933–
Great geological controversies / A. Hallam.—2nd ed.
Includes bibliographical references.
1. Geology––History.   I. Title.
QE11.H35   1989   550'.9—dc20   89–36991
ISBN 0–19–858218–8
ISBN 0–19–858219–6 (pbk.)

9 8 7 6 5 4 3 2 1
Printed in the United States of America
on acid-free paper

Preface to the second edition

My initial response to the publisher's invitation to write a second edition with an additional chapter on the recent controversy about mass extinctions was one of apprehension. I pointed out that, unlike the other controversies reviewed on the first edition, this one was by no means resolved; it is still the subject of active research. Furthermore I had a personal involvement. Having been reassured that, such was the general interest engendered that even a brief account would be welcome, I accepted the invitation, encouraged by the fact that acquaintance with most of the leading participants was an advantage to counteract the risk of offering a partial view, which I have in the event striven to the utmost to avoid. In the circumstances, what I have written can be considered to be no more than an interim report of the situation as I see it in late 1988, and is no substitute for the much more detailed account by Bill Glen, based on hundreds of tape-recorded interviews, that one hopes will be published eventually.

I have taken the opportunity of making a number of additions and amendments to the other chapters, and to add a chapter on two major stratigraphic controversies that took place last century. This has been made possible by the recent publication of two excellent books by Rudwick and Secord. I have attempted to place the controversies in the broader context of work on erecting the stratigraphic column, which is one of geology's most triumphant and fundamental achievements.

With regard to the mass extinctions controversy I have benefited from discussions with European and American scientists too numerous to be mentioned here, but I should like to single out with special thanks for their kindness and hospitality Chuck Officer and Frank Asaro.

Birmingham   A.H.
December 1988

Preface

We have come a long way from acceptance of the conventional Victorian belief in the disinterested scientist, engaged in the objective pursuit of Truth, to a less lofty but more realistic one which takes account of the existence of a whole range of social interactions within the scientific community as determinants of scientific theory. Quite apart from the substantive issues involved, conflicts may arise as a result of rivalries between individuals and research groups. Schools of thought related to institutions, or even nations, can play a role, and prejudice for one reason or another may frequently complicate the straightforward interpretation of factual data.

If one agrees that the essential characteristic of scientific activity is the critical evaluation and testing of conceptual models, then argument can be seen as absolutely integral to it, and the roles adopted by individual scientists in a given argument may well be conditioned to a considerable extent by their aptitude and temperament. I like to classify scientists that I know into arm-wavers and nit-pickers – arm-wavers are better at dreaming up ideas, nit-pickers better at finding fault with them. Most of us, of course, are something of a mixture. Perhaps it is not too cynical to suggest that we tend to be arm-wavers when assessing our own ideas and nit-pickers in evaluating those of our colleagues, but we all recognize the value of fruitful interaction between the two types. It is the discipline of this constant recourse to critical testing of ideas and assessment of data, rather than any inherent nobility, that keeps scientists comparatively honest.

I have undertaken this book with two objectives in mind. In the first place, a number of models of scientific method and progress have been widely discussed in recent years but, in my view, insufficiently tested with reference to the historical development of particular sciences. Since the history of geology is wide open to this kind of analysis, I have chosen eight case histories from celebrated controversies extending back to the very birth of the discipline. Focusing on controversy is helpful because issues tend to get dramatized and the underlying assumptions and attitudes of the protagonists often brought out into the open. Furthermore, attention is concentrated on the matters most critical to growth and development of a given subject.

The second objective is to contribute towards the transmission of some fascinating pieces of intellectual history to a wider audience. I have frequently been dismayed during my years as a university teacher by the common attitude of students, and even some colleagues, who know little and care less about how the modern subject of geology has come about. This state of affairs leaves much to be desired when, for example, even a matter as fundamental as Lyell's uniformitarian doctrine is so often glibly reduced to the triteness of 'The present is the key to the past'. This is about as adequate as rendering Darwin's theory of evolution by natural selection by the phrase 'The survival of the fittest'.

Although this book makes no claim to giving comprehensive coverage to the history of geology over the last couple of centuries it does deal with many of the most important changes in geological thought during that time. No other book brings together the various subjects under review. What might be called the standard textbooks of the history of geology, by Adams, Geikie, and Zittel, cover only the ground of the first two chapters and were written under the influence of an outmoded historiographic tradition. In recent years there has been much critical re-evaluation of the conflicts described, beginning with Gillispie's *Genesis and geology*, which until now has for the most part reached only a limited audience of specialists in the history of science. I have already written at length on the subject of Chapter 6, continental drift, but I have tried as far as possible to avoid repetition, and have introduced many new quotations.

Quotations in fact occupy a substantial part of the text, because I think that well-chosen examples bring a subject to life more than would endless paraphrasing. Some of the quotations rank as minor literary gems, because geological writing in earlier times often exhibited more eloquence and racy phraseology than is customary nowadays. Besides some of the more important pronouncements of a number of great (and not so great) men, which deserve to be more in the public domain, I have included comments or anecdotes recorded by contemporaries, if they serve to illuminate a personality or an attitude.

Although the book has been written with professional geologists and geology students in mind I hope it will also prove to be of interest to a wider audience. The level of prior geological understanding required is very modest, and there is only a minimum of jargon. In particular, geographers should have few problems of comprehension, and biologists may find the geological background to Darwin's evolutionary thinking illuminating. Those interested in the history of science may find useful documentation and food for thought in the case histories, which embrace topics of quite general importance.

The first two controversies took place during what has been called the heroic age of geology. Geology was *the* glamorous science in the first few decades of the nineteenth century, and significantly affected ways of thinking about our planet, most notably with regard to the concept of time. Over 3000 people attended a lecture given by Sedgwick at the Newcastle meeting of the British Association in 1838, and no fewer than 4500 applied for tickets to attend the lecture series that Lyell gave in Boston on his American visit in 1841. Bearing in mind the much smaller educated public, in both relative and absolute terms, we obtain from such figures some idea of the enormous popular appeal of the subject in those days. A knowledge of at least the early history of geology should be considered a basic requirement of anyone with pretensions to being well informed about the way Man has thought about his environment.

In conclusion I should like to record my indebtedness to Dr N. Rupke for his critical reading of the manuscript, Prof. M. Ruse and Drs H. Frankel, W. Glen, and M. J. S. Hodge for illuminating discussions, and Miss H. Gibbs for her efficient typing.

Birmingham A. H.
March 1982

Contents

A theory which is founded on a new principle, a theory which has to make its way in the public mind by overturning the opinions commonly received by philosophising men, and one which has nothing to recommend it but the truth of its principles ... must meet with the strongest opposition from the prejudices of the learned, and from the superstition of those who judge not for themselves in forming their notions, but look up to men of science for their authority...

In order, then, to obtain the approbation of the public, it may not be enough to give a theory that should be true, or altogether unexceptionable; it may be necessary to defend every point that should be thought exceptional by other theorists, and to show the fallacy of every learned objection that may be made against it. It is thus, in general, that truth and error are forced to struggle together, in the progress of science; and it is only in proportion as science removes erroneous conceptions, which are necessarily in the constitution of human knowledge, that truth will find itself established in natural philosophy.

JAMES HUTTON
Theory of the Earth (1975)

1
Neptunists, vulcanists, and plutonists

It is not unusual for the creators of a new discipline within science to deny their intellectual ancestors, rather as most adolescents rebel at some stage against their parents. No doubt this is a necessary phase of growing up and should be viewed indulgently. What is extraordinary about the self-proclaimed founders of modern geology, from Werner to Playfair and Lyell, is that until very recently standard historiography has taken them at their word and accepted their contemptuous dismissal of earlier thinkers about the earth as mere undisciplined speculators. Geology, according to Lyell, was not to be confounded with cosmogeny.

The term cosmogeny refers to the many theories about the earth put forward at the turn of the seventeenth and eighteenth centuries, and continuing well into the eighteenth century with the work of Buffon. As Porter[1] has clearly established, many of the theorists in question, far from having exhibited what Playfair termed 'a species of mental derangement', were counted amongst the leading scientific intelligentsia of the day. Thus the scientific distinction of such continental cosmogenists as Leibnitz and Buffon has long been recognized, as has that of the English scholars Halley and Hooke. Furthermore, it is instructive to learn from Porter's book that Whiston's work was much admired by Locke, and Burnet's by Newton. Thomas Burnet's *Sacred theory of the earth* has, however, been generally dismissed by subsequent generations as mere fantasy.[2] Nevertheless, Gould[3] makes the good point that, however fantastic Burnet's ideas may seem today, he showed originality for his time in seeking natural causes rather than divine intervention to explain how a perfect paradisal planet with a smooth and featureless surface had been severely disturbed by the Great Flood; in fact he was as much a deist as one of his leading critics, Hutton.

The creation of a scientific 'myth' about the earth as a whole was an important advance on an earlier tradition that treated various mineral objects and geomorphological features in isolation, and the fact that less emphasis than later was placed on empirical verification does not detract from the significance of this phase in the growth of geology. Theories remained teleological and anthropocentric, but under the influence of Newtonian mechanistic philosophy the emphasis gradually shifted from a planet in decay to one which was essentially stable and ordered, whether viewed as behaving passively or as a system involving a dynamic equilibrium or forces.

A reaction against vague and grandiose speculation spread throughout Europe in the course of the eighteenth century and an emphasis began to be placed on careful and detailed observation, with little inclination to generalize. Ironically, the first major contribution to establishing geology as a new science, which involved substantial generalization from limited observations, came from one of the principal espousers of this rigorously empirical approach.

The Wernerian doctrine

Abraham Gottlob Werner[4] (1749–1817) was born in Prussian Silesia and from childhood had a close association with mining, making a hobby of collecting mineral specimens. He studied at the newly-established Mining Academy at Freiberg in Saxony, situated close to the rich mining region of the Erzgebirge. After a period of further study at the University of Leipzig he was appointed to a teaching position in the Freiberg Mining Academy in 1775, a position he was to hold for 40 years.

Werner did not enjoy good health during much of his adult life and this might be a major reason for his doing only a limited amount of fieldwork, confined to the Erzgebirge and neighbouring parts of Saxony and Bohemia. He also had a great distaste for writing; hence, besides being a poor correspondent, his publications are very modest both in quantity and length (he would never have got tenure in a modern American university). It was as a teacher that he achieved his enormous fame, attracting students from all over Europe and sending them away fired with enthusiasm.

He prided himself on his capacity for orderliness and methodical study. These were characteristics very well suited to bringing some system into the chaos of mineral and rock classification, a subject to which he made a lasting contribution. But he went far beyond mere classification of minerals and rocks and the more general, synthetic, part of his teaching concerned what he called *geognosy*, a term proposed in the mid-eighteenth century by Füchsel. Geognosy was defined by Werner as the 'science which treats of the solid body of the earth as a whole and of the different occurrences of minerals and rocks of which it is composed and of the origin of these and their relations to one another'.

Werner's mineralogy teaching embraced more than just mineral substances. In his lecture classes he would expand from a consideration of individual specimens to physiography and human affairs, pointing out how the arts and industry were influenced by minerals and rocks. According to Cuvier

He treated his subjects in such an admirable manner that he roused the enthusiasm of his hearers and inspired them not only with a taste but with a passion for his science.... At the little Academy of Freiberg, founded for the purpose of training mining engineers and mine captains for the mines of Saxony, there was renewed the spectacle presented by the universities of the Middle Ages, students flocked thither from every civilized country. One saw men from the most remote countries, already well advanced in years, men of education holding important positions, engaged with all diligence in the study of the German language, that they might fit themselves to sit at the feet of this 'Great oracle of the sciences of the earth'.[5]

As Werner published very little, a full account of his geognostic teaching has to be gleaned partly from his unpublished manuscripts and the lecture notes of his students. The most complete statement in English was published by Werner's leading British disciple Robert Jameson.[6] Werner's key publication was a mere 28-page pamphlet, the celebrated *Kurze Klassifikation und Beschreibung der verschiedenen Gebirgsarten*, which was completed in 1777 but not published until a decade later.[7] In this he outlined what was in effect a stratigraphic scheme that was claimed to be applicable to the whole earth. Initially, in the *Kurze Klassifikation*, there were only four stratigraphic units (*Gebirge*, an old mining term) but subsequently a fifth, *Ueber-gangsgebirge*, was added. In order of decreasing age the units are as follows:

    1. *Urgebirge* ('Primitive strata'). Granite, gneiss, schist, serpentine, quartz, porphyry, etc.;

    2. *Uebergangsgebirge* ('Transitional strata'). A succession, now attributed to the late Palaeozic, of limestone, diabase, and greywackes;

    3. *Flötzgebirge* (Floetz strata). Twelve subdivisions ranging in succession from what is now known as Rotliegendes, Kupferschiefer, and salt deposits (all Permian), Triassic, Jurassic, Upper Chalk, and Tertiary, which includes brown coal and basalt. The relative age of the Upper Chalk with respect to the Jurassic and Tertiary was not clearly established;

    4. *Aufgeschwemmte Gebirge* ('Swept together strata'). Relatively unconsolidated deposits (Nagelfluh conglomerate, sands, clays, etc.);

    5. *Vulkanische Gesteine*. Both true volcanics (lavas, tuffs) and 'pseudovolcanics' (hornstone, porcelain jasper).

Werner's stratigraphic scheme was not original. It owes much to the work of his German compatriots Lehman and Füchsel who, by the mid-eighteenth century, had established the main features of the stratigraphic successions in the Harz Mountains, the Erzgebirge, and Thuringia. The term Flötzgebirge comes from Lehman, while Füchsel, who records more details of the stratigraphic succession, was the first to distinguish actual rock formations such as the Kupferschiefer, Zechstein, and Muschelkalk. Werner was most fulsome in acknowledging his debt to the Swede, Torbern Bergman, who in turn frequently quoted Lehman without mentioning his quite similar and earlier published *Uranfängliche* and *Flötzgebirge*.[8]

Not only did Werner add petrological details to the earlier proposed successions, he generalized them into a global scheme and proposed a theory to account for their origin, and in this lay his claim to originality. Little was known of the inner part or 'nucleus' of the earth but it evidently had an irregular surface, with high elevations alternating with deep depressions. Initially the earth was completely enveloped by a primeval ocean[9] covering even the highest mountains. Deep, turbid waters held in suspension or solution all the materials that now form the crust. As time passed a succession of deposits was laid down. Initially these deposits were exclusively chemical precipitates (and included granite and other rocks we now accept as igneous) with gneiss and schists, which together comprise the Primitive Rocks. As the waters began to subside, rock formations were laid down which consisted partly of chemical precipitates and partly of mechanically deposited sediments (Transitional Strata). Rare marine fossils occurred. With a further drop in ocean level the Floetz strata were deposited, with mechanical deposits coming to predominate over chemical precipitates, and with fossils often occurring in abundance. Finally alluvial deposits were laid down in lowland areas, having therefore only a local distribution; they were derived mainly from mechanical disintegration of older rocks.

In the Primitive and Transitional formations strata are often steeply inclined. This was partly because they were chemical precipitates; it was well known that when a substance crystallizes out from solution it will deposit on the sides as well as the bottom of the containing vessel. Hence the steep inclination of the strata could indicate their conforming to the original contours. Steep inclinations were also attributed, however, to the irregular manner in which uncompacted deposits have settled down, and to masses of precipitates having slipped down steeply inclined surfaces. The primitive ocean in fact did not subside slowly and quietly but was very turbulent, with powerful currents cutting deep channels to produce valleys and mountains. As the ocean water became progressively quieter through time so the strata tended towards horizontality and had a progressively more restricted distribution as water level dropped.

The emphasis upon the role of water in the theory quickly led to its being dubbed *neptunist*. The dogmatic tone adopted by Werner is brought out clearly in the following passage from another of his publications, on the origin of veins, which were also thought to be produced by chemical precipitation from the universal ocean, and included crystalline rocks as well as minerals.[10]

In recapitulating the state of our present knowledge it is obvious that we know with certainty that the floetz and primitive mountains have been produced by a series of precipitations and depositions formed in succession from water which covered the globe. We are also certain that the fossils [i.e. minerals] which constitute the beds and strata of mountains were dissolved in this universal water and were precipitated from it; consequently the metals and minerals found in primitive rocks, and in the beds of floetz mountains, were also contained in this universal solvent, and were formed from it by precipitation. We are still further certain that at different periods, different fossils have been formed from it, at one time earthy, at another metallic minerals, at a third time some other fossils. We know, too, from the position of these fossils, one above another, to determine with the utmost precision which are the oldest, and which are the newest precipitates.

On the other hand there are contemporary reports which throw some doubt on the extent of Werner's dogmatism, which was perhaps no greater than that of most German professors of the period. Thus Pinkerton[11] reports Werner as saying that 'a theory is useful to concentrate facts, and render them more clear and pleasing to an audience'. 'Nor', continues Pinkerton, 'with the modesty of a man of real genius, did he conclude his own theory to be unobjectionable.' It is evident at any rate that Werner was prepared to amend his theory without abandoning the fundamentals. We have already seen that he added the transitional strata after publishing his *Kurze Klassifikation*. When he learned that some 'precipitates' occurred out of due sequence and at much higher altitudes than previously supposed, he was prepared to envisage a temporary return of his universal ocean to rise again over the hills from which it had earlier receded. Thus over the area covered by Floetz deposits the waters precipitated basalt and then descended.

To state that Werner's theory had great appeal would be an understatement. It appeared to account in a simple, satisfying way for a wide range of geological phenomena; it successfully bridged the gap between mere cosmogenic theories like Buffon's, which postulated a series of episodes in earth history without providing supporting evidence, and the numerous but scattered sets of empirical observations by diligent scholars in several countries of Europe. Saussure in the Alps and Pallas in the Urals had confirmed the fact first established in Central Europe that the cores of mountain ranges consisted of crystalline rocks, evidently older than the strata that overlay them, and also formed a basement in lowland areas. The proportion of detrital (i.e. mechanically deposited) to chemically precipitated (limestones, evaporites) sediments of the younger formations appeared to increase through the stratal time sequence, and alluvial deposits were restricted in distribution to lowland areas. The older the strata, the more contorted they tended to be, suggesting earlier times of great turbulence.

No wonder, given also Werner's charismatic personality as a teacher, that he produced a succession of enthusiastic proselytising neptunist disciples anxious both to spread the gospel and explore different parts of the globe to elucidate the Wernerian succession. Nevertheless there was no shortage of contemporary sceptics. One of the most influential of Werner's critics was the Italian geologist Scipio Breislak.[12] In his much used textbook of geology he argued that the volume of water now present on the globe was utterly insufficient to contain in solution or suspension all the solid material of the crust. So where had all the water gone? The Wernerians were never able to give a satisfactory answer to this question, and Werner himself seems to have considered that it disappeared for some unknown reason into outer space! Breislak, along with others, thought that many of the geological phenomena in question could better be explained by advocating uplift of the land rather than fall of ocean level. As an Italian Breislak was inevitably impressed, like Moro many years before him, with the power of volcanoes and their capacity to cause uplift, while as early as the seventeenth century Steno had treated the inclined and broken strata of northern Italy as evidence of crustal disruptions. This pioneer Italian work was totally ignored by Werner.

The major conflict was, however, to arise over another subject, the origin of basalt. Was it, as the neptunists argued, a chemical precipitate from the universal ocean, or was it a lava erupted from volcanoes, as proposed by a group of mainly French and Italian geologists who became collectively known as vulcanists?

The origin of basalt

The name basalt had been resurrected in the sixteenth century by Agricola, who had rediscovered Pliny's ancient term for a distinctive dark-coloured, crystalline rock, often exhibiting columnar jointing, that was widely recognized by geologists in the latter part of the eighteenth century. In Werner's native Saxony and elsewhere in

Central Europe, horizontal or subhorizontal layers of basalt formed a capping to certain hills, notably in his Floetz formation, while the evidently related greenstones occurred in his Primitive and Transition formations. While Werner was perfectly well aware that others considered basalt to be volcanic in origin he disregarded their views without attempting a detailed refutation and persistently maintained that it was a chemical precipitate from the universal ocean. The highly inclined sheets of basaltic rock obliquely cutting across the strata and known as dykes or trap rock gave him no trouble. They came into his broad category of veins which were evidently all fissure fills composed of chemical precipitates.

The existence of volcanoes could not, of course, be denied, but Werner restricted their activity to very recent times. Since there was no 'interior fire' to serve as a source of heat, Werner argued that eruption of lava took place where basalt and other rocks were melted by the combustion of underlying seams of coal. Coal was known from both Transition and Floetz strata in the Germanic countries and Werner and his students spent much time seeking to establish a close relationship between the distribution of basalt and coal. Werner had no direct field experience of volcanoes but he had of course seen samples of lava. So how did he account for the strong resemblance of lava and basalt? He argued that, where basalt was found in the condition of a lava, the original precipitate had been fused by fire produced by the combustion of underlying coal deposits. Lavas were often associated with scoriae and volcanic cones and had no columnar jointing. The basalt layers that Werner was familiar with had, on the other hand, no scoriae but possessed columnar jointing, and were intercalated in sequences of sedimentary strata indubitably of aqueous origin. He had made observations in 1777 concerning an old subterranean fire in a coalfield surrounding basaltic hills in Bohemia. Rocks altered by such fire included baked clay ('porcelain jasper') and were grouped together as pseudovolcanic.

It was not a new idea that the presence of combustible substances such as coal and bitumen was all that was required to account for the production of volcanoes, for it had been proposed in mid-century by the French geologist, Jean Etienne Guettard (1715–86). Unlike his illustrious contemporary Buffon, Guettard was an empiricist who did valuable pioneer work in geological mapping and geomorphology. Most important in the present context, however, was that he was the first to recognize from careful observations the existence of extinct volcanoes in the Auvergne district of the Massif Central, which were to figure so largely in the later controversy. Guettard did not, however, think that basalt was a volcanic rock but believed instead that it formed by crystallization from an aqueous fluid. Hence in a curious way Guettard can be regarded as the parent of both the vulcanist and neptunist schools of thought.

A far better claim to being the father of the vulcanist school can be made for another Frenchman, Nicholas Desmarest[13] (1725–1815), whose research in the Auvergne had wide influence in France. Unlike Werner, Desmarest was not an academic but a government employee who eventually rose to become Inspector General and Director of French manufacturing industry. His intellectual distinction was widely acknowledged during his lifetime, as recognized by his being invited to contribute to the celebrated Encyclopaedia of Diderot and d'Alembert. His geological interest was fanatical and single-minded, and he was positively obsessed by rocks to the exclusion of much else, as illustrated in the following extract from Cuvier's *éloge*.[12]

His friends used jocularly to affirm that he would have broken the most beautiful statue in order to ascertain the nature of an antique stone, and his character was so widely given to him that at Rome the keepers of the museums felt some alarm in admitting him. In society, too, things, whatever they might be, affected him on one side only. For instance, when an Englishman was recounting at the house of Duchesse d'Anville the then recent thrilling incident in Cook's first voyage, when his vessel, pierced by a joint of rock, was only saved from sinking by the stone breaking off and remaining fixed in the hole, everyone present expressed in his own way the interest he felt in the story. Desmarest, however, quietly inquired whether the rock was basaltic or calcareous.

Like Guettard, Desmarest was staunchly empirical in his outlook and his work in the Auvergne is an outstanding example of painstaking, detailed research in one small area with bearing on an important scientific problem. In 1763 he detected columnar jointing in volcanic rocks which he was able to relate to an old lava stream, and also traced similar rocks with fresh scoriae into old craters. He was well aware of the nature of the columnar jointing in the Giant's Causeway in Northern Ireland, which became widely known to European naturalists in the eighteenth century. Though he never saw the Giant's Causeway, he did pay a visit to Italy to examine active volcanic phenomena and also returned to the Auvergne on more than one occasion. Desmarest was the first to recognize the significance of outliers and the removal of scoriae and slags as evidence of isolation and destruction by erosion, with craters being progressively eliminated. These provided the first concrete examples of the doctrine of origin of valleys by the erosive action of streams, and the first attempt to trace back the history of a landscape by comparing different erosional stages.

Desmarest published what was quickly to be recognized as a classic monograph in 1774, as a memoir of the Academy of Sciences. As might be expected of an empiricist, he tended to be very cautious and restrictive in his interpretations but he allowed himself the indulgence of speculating that Guettard was wrong in attributing the heat source for vulcanism to coal and bitumen, preferring to believe that the melting of granite could produce basalt!

Further evidence of the igneous origin of basalt in the Massif Central (Auvergne, Vivarais, and Velay) was brought forward in the succeeding years by Soulavie, Faujas de Saint-Fond and Dolomieu, and other confirmatory papers were written by Raspe of Hesse and Arduino of Padua. Arduino's compatriot Breislak was of course a staunch vulcanist.[15] The criticisms of the vulcanists, never openly met in publication by Werner, seem nevertheless to have provoked him to seek further evidence for his neptunist views. He thought he had found the critical evidence required in 1788, on the Saxon hill known as Scheibenberg, where he described an upward sequence of sandstones, clay, 'wacke' (= mudstone), and basalt which appeared to grade into each other.[16] The 'perfect' transition clearly indicated to him that basalt could not be distinguished from sediments, and that therefore an aqueous origin was undeniable. He triumphantly challenged the vulcanists to explain away what he evidently considered documentation for his ideas.

Unfortunately for Werner this new 'discovery' was quickly challenged by one of his most able and favourite students, J. K. W. Voigt, who insisted that the Scheibenberg basalt was an ancient lava. A lengthy controversy ensued, which converted neither party but resulted instead in the rupturing of a beautiful friendship!

In fairness to the Wernerians, Desmarest admitted that, confronted with the evidence of the Saxon Hills alone, he could not have determined that basalt is volcanic in origin. He otherwise kept aloof from the neptunist-vulcanist controversy, which progressively increased in momentum in the closing years of the century, and he simply exhorted the volcanic sceptics to go and see the evidence in the Auvergne for themselves.

The controversy was soon to be expanded in a way that would strike even more fundamentally at the Wernerian doctrine. On one point neptunists and vulcanists were agreed. Granite was primordial rock, part of the original crust. Now it had been known for some time that the Auvergne volcanic rocks overlie granite. In 1789 Guy de Dolomieu (1750–1801), Professor at the École des Mines in Paris, matched socio-political events in his country by putting forward the revolutionary proposal that the granite was not primordial but was underlain by rock of very different composition, which had penetrated the granite to give rise to basaltic lava.[17] The volcanic hearth could not therefore be located in sedimentary strata containing combustible materials, and the heat source must lie at some considerable depth below the consolidated crust. Contemporary research by James Hutton in Scotland was to bring forward evidence that granite itself was igneous in origin, and in many cases had penetrated overlying sedimentary strata. Fire rather than water might be the key to a wide range of geological phenomena.

The plutonism of Hutton

James Hutton[18] (1726–97) was interested in chemistry from the days of his youth and chose to study medicine at Edinburgh University as the subject most allied to this interest. Subsequently he studied at the Sorbonne in Paris and Leiden University in Holland, where he obtained his doctorate for a thesis on blood circulation. He never took up medicine as a profession, however. Instead he became a farmer in Berwickshire in 1754 after spending two years in East Anglia. His interest in chemistry persisted and he became a successful partner in a small industrial concern involving the manufacture of sal ammoniac. This provided him with the financial means to abandon farming and return to Edinburgh in 1768 as a gentleman of leisure, who could devote himself to scholarly pursuits, and there he was to stay for the rest of his life. Like Werner a lifelong bachelor, his social life revolved around the Oyster Club, which included such close friends as Joseph Black the chemist, John Playfair the mathematician, and John Clerk of Eldin the naval tactician, together with the economist Adam Smith, the architect Robert Adam, and the philosophers Adam Ferguson and Dugald Stewart. With the great philosopher David Hume also living in Edinburgh, it was not for nothing that the city was known at the time as the Athens of the North. Paris apart, it was without peer as a centre of scholarly and intellectual talent.

Of all these men it was Black who had the greatest scientific influence on Hutton. Their scientific attitudes and personalities could hardly have been more different, according to Playfair.[18] 'Hutton's ardour, enthusiasm, rapid thought and animation were met by Black's caution and coolness. Hutton's dread of ignorance, by Black's fear of error, Hutton's curiosity was imperious, but Black's could be laid aside, Hutton's simplicity was careless and often in collision with popular prejudice, whereas Black's was correct and respected popular prejudice.'

Hutton indeed appears to have been an attractive and stimulating companion who would surely have made a great teacher, perhaps comparable to Werner. Playfair records that 'The fire of his expression, on such occasions [other people's discourses], and the animation of his countenance and manner are not to be described, they were always seen with great delight by those who could enter into his sentiments, and often with great astonishment by those who could not'. As a scientist he combined to an unusual degree a skill in acute observation with a capacity for bold and original theorizing.

As in the case of Darwin there was a long gestation period before he published his great work but, unlike Darwin, he died shortly afterwards. It appears likely that he had decided on the igneous origin of both basalt and granite as early as the mid-1760s and formulated his basic theory shortly afterwards. He evidently had to be prodded by his friends into making his work public and eventually read a paper to the Royal Society of Edinburgh in 1785. Three years later this was published in volume one of the Society's *Transactions* with the cumbersome title: *Theory of the Earth; or an investigation of the laws observable in the composition, dissolution, and restoration of land upon the globe.* The style of writing is a strange one for the modern reader. Even by the standards of contemporaries it is steeped to an unusual degree with a teleological outlook, with frequent expression of the quite conventional eighteenth (and also to a considerable extent nineteenth) century belief in a grand design or purpose in nature. Thus: 'We perceive a fabric, erected in wisdom, to obtain a purpose worthy of the power that is apparent in the production of it'. This purpose is to maintain the earth as 'a habitable globe'. We 'are led to acknowledge an order, not unworthy of Divine wisdom'.

It is not surprising to read therefore that 'A volcano is not made on purpose to frighten superstitious people into fits of piety and devotion, nor to overwhelm devoted cities with destruction; a volcano should be considered as a spiracle to the subterranean furnace, in order to prevent the unnecessary elevation of land, and fatal effects of earthquakes'. In other words it acts as a kind of safety valve, an idea that can be traced all the way back to Strabo. Hutton classified the igneous rocks in the Auvergne and the Eifel as 'proper lavas' and the Scottish 'Whinstones' (i.e. basalts or dolerites) as 'subterraneous lavas'. By this he meant that they were intrusive rather than extrusive. Since he considered granite also to be an intrusive igneous rock, occurring in bodies of huge size compared to whinstone sills and dykes, and stressed the importance of subterranean heat in geological processes, Hutton and his supporters acquired the epithet of plutonist.

The Midland Valley of Scotland contains numerous basic sills of

Carboniferous age, and it was on one of these, Salisbury Crags in his native Edinburgh, that he made the critical observations that apparently first convinced him of the igneous and intrusive origin of whinstone, notably of chilled margins and baked country rock. He never was persuaded of the existence in Great Britain of 'proper lavas' as could be found on the continent, though plenty were subsequently recognized.

Shortly afterwards the igneous concept was extended to granite, based essentially on the graphic texture, readily observable with the naked eye, of the Portsoy granite in north east Scotland. He inferred correctly that the quartz and feldspar crystals must have crystallized simultaneously from a state of fusion.

To prove his plutonic thesis he needed to demonstrate that the granite had not resulted from fusion of pre-existing sediment *in situ* but had been intruded upwards, as had whinstones. Oddly enough he had not succeeded in doing this when he presented his theory in 1785, but later that same year he and Clerk of Eldin decided to make an attempt. They chose an area where granite and schist were known to be in contact, on the Duke of Atholl's estate in Glen Tilt in the southern Grampian Highlands. Let Playfair's words tell the story.

When they had reached Forest Lodge, about seven miles up the valley, Dr Hutton already found himself in the midst of the objects which he wished to examine. In the bed of the river many veins of red granite (no less, indeed, than six large veins in the course of a mile), were seen traversing the black micaceous schistus, and producing by the contrast of colour, an effect that might be striking even to an unskillful observer. The sight of objects which verified at once many important conclusions in his system, filled him with delight; and as his feelings, on such occasions, were always strongly expressed, the guides who accompanied him, were convinced that it must be nothing less than the discovery of a vein of silver or gold, that would call forth such strong marks of joy and exultation.

In the next two years, Hutton made further discoveries in Galloway and Arran of intrusive granite veins, so that at last he had found conclusive evidence that some granite at least was younger than the enveloping country rock and hence could not be the primordial material of common belief.

Although elements of his ideas were derivative like Werner's the thesis Hutton put forward in his 1788 paper was almost startlingly original. Many years earlier the Venetian Lazzaro Moro (1687–1740) had been deeply impressed by the comparatively recent volcanic eruptions of Santorini in the Aegean Sea and the formation of Monte Nuovo in the Phlegrean Fields close to the Bay of Naples, and studied descriptions of similar phenomena in the ancient world by Pliny and

Strabo. This led Moro to extrapolate to mountains and propose that their uplift was due to an uprush of fiery gases and lava from the earth's interior. Furthermore, a good deal was already known about denudation and sedimentation and the actualist principle of studying present processes to understand the past had already been practised in a sporadic way by a number of people.

The originality of Hutton lay in his formulating a cyclic steady-state model of the earth, which he visualized to exist in a condition of dynamic equilibrium, the beginning and end of whose immensely long history was in principle unknowable. The buildup of heat from the earth's interior, periodically relieved by vulcanicity, caused the uplift of land, provoking increased erosion to wear it down. The resulting sediments were deposited in the sea and subsequently consolidated. The agent of consolidation could not be water, because of the insoluble nature of sedimentary rock cements. Hutton was here influenced by Black's experiments which appeared to suggest that a combination of heat and pressure with depth would lead to consolidation.

The expansive power of this interior heat would lead eventually to uplift of the sea bed, as evinced by the presence of marine fossils on land, in turn to be destroyed by denudation, and so on in a series of cycles over an indefinitely long period of time.

Intrusive rocks, categorized as whinstone, granite, and porphyry (including felsite and quartz porphyry), were the most telling witnesses to the importance of heat. Hutton's plutonism was therefore much more fundamentally opposed to Werner's passive-earth model exhibiting progressive change than ever the views of the vulcanists were. Desmarest denied an igneous origin to granite until his death, but as staunch a neptunist as de Luc was quite prepared to accept that basalts were volcanic rocks.

Consequently one might have expected a massive critical onslaught from the Wernerians but in the event Hutton's paper was largely ignored, except for attacks by de Luc and Kirwan, principally directed at Hutton's challenge to the widely accepted Mosaic chronology, which will be discussed in the next chapter. Kirwan also challenged the igneous interpretation of granite. His attack[19] was couched in such bitter terms that Hutton was stung into writing a much more fully documented account of his theory ('with proofs and illustrations') which appeared in 1795, two years before his death, as a two-volume treatise.[20] The manuscript of a third volume was discovered a century later and published by the Geological Society of London in 1899.

The first chapter of Volume 1 is an almost word-for-word

reproduction of the 1788 paper and amazingly fails to mention the results of his great explorations in 1785–8, when he discovered not only intrusive granite veins but angular unconformities.[21] Chapter 2 is a response to Kirwan's criticism. With regard to Kirwan's claim that granite is an aqueous product, Hutton contemptuously dismisses his cited example of a mole across the River Oder, pointing out that there was no evidence for Kirwan's extraordinary claim that granite rock had been produced by passing water through granitic sand. It was far more plausible to maintain that the unconsolidated sand had been clogged by mud! Remarkably enough Hutton fails to mention his evidence from Glen Tilt, Galloway, and Arran supporting his ideas on the intrusive origin of granite.

In Chapter 3 the neptunists are criticized (Werner is not mentioned by name) but so are the vulcanists for failing to recognize the connection between volcanoes and the elevation of land, together with their readiness to accept the amygdales in basalts as due to infiltration into vesicles (which is indeed the correct explanation). The next chapter is concerned with a challenge to the notion that the so-called Primitive rocks exposed in the cores of mountain ranges are part of the original 'nucleus' of the globe, formed prior to all 'organized matter' (i.e. fossils), and records that such rocks, as for example in the Lake District and the Scottish Southern Uplands, have been found to contain both pebbles, implying erosional and sedimentational processes, and fossils. It is here apparent that Hutton is attacking the interpretation of Lehman and his contemporaries, although no names are mentioned, and he was evidently unaware that Werner had transferred such rocks from the Primitive to the Transitional formation, which weakened the force of Hutton's challenge. Hutton goes on to discuss the origin of granite, regarded by both neptunists and vulcanists as Primitive. (The discoveries in Glen Tilt, Galloway, and Arran were not referred to until the posthumously published Volume 3.)

The only other part of Volume 1 that need concern us here is Hutton's remarks in Chapter 6 on his discovery of angular unconformities in Arran and at Siccar Point, Berwickshire. Hutton was the first to appreciate the significance of such phenomena as indicating a historical sequence: sediment deposition and consolidation – tilting and uplift – erosion – sediment deposition and consolidation.

Discussion of the evidence for elevation continues in Chapter 1 of the second volume, a substantial part of which is devoted to Saussure's pioneer descriptive work of the Alps, on which Hutton depended entirely for his knowledge of that great mountain range.[22] The most critical evidence, recorded in Volume 2 of Saussure's work published in 1786, concerns his discovery at Valorsine of a thick conglomerate consisting of interbedded layers of pebbles and sand oriented vertically. For a long time Saussure, who adopted Wernerian views, had believed that vertical or steeply dipping strata flanking Alpine granites occurred in their original position of formation. The Valorsine conglomerate persuaded him otherwise. He was forced to admit that it must have been laid down horizontally and subsequently disturbed by earth movements.

Hutton was evidently elated by this discovery because it challenged a fundamental tenet of the Wernerian doctrine. It was widely recognized that bands of schist commonly had steep dips but because of the crystalline nature of these rocks there was ambiguity about their interpretation and it was not self-evident that Hutton's explanation – that they were original sediments altered by heat – was correct. About an unaltered rudaceous rock such as the Valorsine conglomerate there could, however, be no doubt to any reasonable observer. It provided unequivocal support for the sporadic recognition since the time of Steno that sedimentary strata could be dislocated and tilted as well as uplifted. Furthermore, on Occam's principle, the long-recognised presence of marine fossils in mountain ranges was more economically explained by uplift from the sea bed than by subsidence of a universal ocean, the progressive disappearance of whose waters remained an unfathomable mystery.

It would have been even more exciting from Hutton's point of view if evidence of volcanic action could be found in the Alps and he was evidently disappointed not to find any such evidence even in the third and fourth volumes of *Voyages* which reached him shortly before his death.

Hutton's great work had little direct impact on the geological community, principally it appears because of his somewhat verbose and obscure prose style and lack of a coherently organized structure, together with his failure to publish the critical evidence for the intrusive character of granite. Such impact had to await a far more lucid account by Playfair at the beginning of the nineteenth century. Meanwhile we must turn back to the findings of some of Werner's leading European disciples.

The researches of d'Aubuisson and Buch[23]

As time passed, increasingly thorough and extensive field investigations demonstrated that the stratigraphic scheme of Werner's *Kurze Klassifikation* required amendment. We have already noted that Werner soon intercalated a Transition formation between his Primitive and Floetz formations. More important because of their bearing on the theory of precipitation from a gradually receding ocean, rocks originally attributed to Primitive rocks, such as basalt and porphyries, were found intercalated in mechanically deposited sedimentary rocks in the Floetz formation, suggesting a return to conditions promoting aqueous precipitation. It was further found that some granite *overlies* slates of the Primitive formations. Such phenomena were descriptively accommodated by referring to, for example, the oldest and newest granites, and to the first, second, third, and fourth porphyry. It was never made entirely clear how these changes were to be explained, and it should have strained the credence of all but the most blindly faithful Wernerians to invoke periodic fluctuations of the level of the universal ocean on such an arbitrary basis. Additionally, there was no ready explanation in Wernerian theory for the considerable variations in thickness of distinctive rock units from area to area.

The most severe blows to the neptunian doctrine came, however, from the defection into the vulcanist camp of three of Werner's most distinguished pupils, who had studied with him in the 1790s, Humboldt, d'Aubuisson, and Buch.

Jean François d'Aubuisson de Voisons (1769–1819) was born in Toulouse and studied at Freiberg from 1797 to 1802. He left Freiberg an ardent neptunist and wrote a treatise on the basalts of Saxony, in which the Wernerian doctrine was treated as incontestable. He was advised by his referees to visit an area where modern volcanoes could be studied and so he went to the Auvergne and the Vivarais. Here he found basalts overlying at least 400 metres thickness of granite. Since *coal* could hardly be supposed to exist under the earliest chemical precipitate these basalts could not be lavas according to Werner's teaching. But he saw craters with lava streams. In fact he was drawn to the same conclusion for the same reasons as Dolomieu some years earlier. He quickly published the results of his investigations[24] but it took some time to acknowledge that the basalts of Saxony must have had a similar origin

The facts which I saw spoke too plainly to be mistaken; the truth revealed itself too clearly before my eyes, so that I must either have absolutely refused the testimony of my senses in not seeking the truth, or that of my conscience in not straightaway making it known. There can be no question that basalts of volcanic origin occur in Auvergne and the Vivarais. There are found in Saxony, and in basaltic districts generally, masses of rock with an exactly similar groundmass, which enclose exactly and exclusively the same crystals, and which have exactly the same structure in the field. There is not merely an analogy, but a complete similarity; and we cannot escape from the conclusion that there has also been an entire identity in formation and origin.[25]

By far the most eminent, however, of the active propagandists for Werner's cause was Leopold von Buch (1774–1853) in the initial years of his brilliant field investigations, which eventually earned him the reputation of being the outstanding geologist of his generation. After leaving Freiberg he worked for a short time for the Mining Service of Silesia before resigning in 1797 to become a full-time researcher. Having been born of good family he was sufficiently well endowed financially to be free to do this. He immediately visited the Alps for the first time and in the following year went to Italy. In the early part of the new century he wrote up his researches in a celebrated travel book dedicated to Werner.[26]

Early in the first volume Buch is quite categoric in his neptunist views. 'Every country and every district where basalt is found furnished evidence directly opposed to all idea that this remarkable rock has been erupted in a molten condition, or still more that each basalt hill marks the site of a volcano.'[27]

It seems likely that the first seeds of doubt were sown after seeing the volcanic rocks around Rome, and on and around Vesuvius and the Phlegrean Fields he was greatly impressed by evidence of the manifest power of volcanic forces. He seems to have been especially bewildered by the discovery of a feldspar porphyry that was indubitably a lava flow, and sought in vain for any evidence of coal deposits to provide the enormous heat source needed. Nevertheless he resisted conversion to vulcanist ideas, but makes no secret of his change of opinion after visiting the Auvergne, that graveyard of neptunist views, in 1802.[28] Like Dolomieu before and d'Aubuisson after him, von Buch was impressed to find volcanoes arising from a plateau underlain by massive granite. He could not deny the volcanic origin of the basalts, nor the plugs containing feldspar porphyry. Nevertheless it is not quite true to assert that he returned to Germany a totally converted vulcanist, because he stubbornly persisted for some time in his belief that the Saxon basalts, interlayered as they were with sediments, were of aqueous origin.

Buch's next major trips, in 1806, 1807, and 1808, were to Scandinavia, and again he published his findings in a popular travel book.[29] With regard to the controversy under discussion, his single most important discovery was in the Kristiana district of Norway, where he observed veins of granite extending up into a fossiliferous limestone whose contact had evidently been altered by heat. He also found and correctly interpreted evidence for the recent uplift of land in various parts of Scandinavia. These results had a significant effect in converting him to the plutonist school, but he never openly acknowledged his debt to Hutton, although it seems unlikely that he remained totally unaware of his work.

Further work in the Canaries, Alps, and elsewhere progressively convinced von Buch of the importance of vulcanicity in generating mountains, but he never formally renounced the Wernerian doctrine, which as we shall see embraced much more than neptunism. The emancipation was a gradual one and, for both him and his contemporaries, neptunist views simply withered away as they became progressively less relevant or informative for current research. Likewise, d'Aubuisson still inclined to the Wernerian system in his treatise published shortly before his death, but rejected parts of it in favour of a plutonist view. Neptunism continued to be taught in German universities for some years after Werner's death but more on grounds of established authority than deep conviction, because it never recovered from the body blows inflicted by the defection of some of Werner's leading disciples. Long after belief in a subsiding universal ocean had faded, widespread use was made of stratigraphic terms like *Transition* and *Floetz*, but by the mid-1820s neptunism was effectively dead. We shall now turn to the well-documented decline of neptunism in the British Isles.

The British controversy

Kirwan returned to his intemperate attack on Hutton in his *Geological essays* (1799) and made great play of his discovery of marine fossils in a basalt layer at Portrush, Northern Ireland, this being considered to be *prima facie* evidence of an aqueous origin. Not long afterwards Hutton's supporters were able to demonstrate that the supposed basalt was in actuality a shale baked hard adjacent to a contact with chilled basalt, though this was never accepted by the more ardent neptunists. One of these supporters was John Playfair (1748–1819), for whom Kirwan's renewed attack was a major stimulus to write a more intelligible and therefore acceptable version of his friend's great work. His *Illustrations of the Huttonian theory*[30] was indeed widely read and quickly became a classic. Not only were Hutton's ideas put forward in far more lucid prose than the master could produce, the providential overtones were played down. New evidence was brought forward in a number of instances, and sometimes he improved on the quality of the arguments. In other words he was more than a mere popularizer.

The other leading plutonist disciple of Hutton was Sir James Hall (1761–1832), who is the pioneer of petrological experimentation. Stimulated by Hutton's ideas on the igneous origin of granite, porphyry, and basalt, he began experiments on the fusion of Scottish and Italian basalt and published his results several years later in a paper entitled *Experiments on whinstone and lava.*[31] He observed that rapid cooling led to the formation of glass whereas slow cooling resulted in a crystalline rock resembling the original material.

In the early part of the nineteenth century Edinburgh was to become the stage of an exceptionally acrimonious controversy between plutonists and neptunists. The leading neptunist was Robert Jameson (1774–1854), who went to Freiberg in 1800 after studying at Edinburgh University. He returned to Edinburgh in 1804 to take up the Regius Chair of Natural History, a post he held for half a century until his death (Playfair held the chair of Natural Philosophy in the same university, having moved from the chair of Mathematics in 1805).

Jameson, like his revered master in Freiberg, was a very capable mineralogist and became Keeper of the University Museum. Volumes 1 and 2 of his *System of mineralogy*, published in 1804, were essentially a descriptive catalogue, but Volume 3 (1808) offers the best contemporary English language description of Werner's doctrine without exhibiting any originality.[32] Jameson stressed, as a good Wernerian should, the virtues of empirical study in the field.

We may remark, that a considerable degree of practical experience in observing nature, and tracing the connections of mineral formations, is required to render us capable of applying the principles of Wernerian geognosy, and even of duly comprehending its value, as a faithful picture of the mineral kingdom. It is also certain, that, if we attempt to write mineralogical descriptions, without having made ourselves acquainted, not only with simple fossils [i.e. minerals] and mountain-rocks, as far as they can be examined in cabinets, but with their mode of aggregation as constituting Mountain Groupes, which can be rightly understood only by studying minutely the seamed and stratified structure of rocks, the various relations of beds and veins, and the characters and connection of the different formations as they appear in nature itself, *on the great scale*, we shall most unquestionably fail in communicating useful information. In such a state of ignorance, we may possibly conceive that we frequently discover marks of dislocation, contortion and confusion, which exist only in our imagination; we may even arrive at the formidable conclusion, that the solid parts of this globe are little else than a heap of ruins. The late Dr. Hutton of this place [Edinburgh], a man of unquestionable ingenuity, but very imperfectly skilled in mineralogy, inferred, from his observations, that the present world has been formed from the debris of two former worlds, inhabited by numerous tribes of animals, and clothed with a profusion of magnificent vegetables. According to this strange hypothesis, even gneiss, mica-slate [mica-schist] and clay-slate are but mechanical deposits, which have been softened by the action of heat, so as to permit their being elevated, without breaking, from their supposed original horizontal position, to their present vertical one.

It is admitted by all geognosts, except the Huttonians, that there exists a great class of rocks, denominated Primitive, which forms the oldest part of the earth, and contains no materials, more ancient than itself, but is a true chemical deposition; consequently that this world cannot be considered as deriving any of its material from one or more worlds, antecedent to it. If the truth of this fact can be established, it is evident, that the Huttonian Theory, notwithstanding the powerful eloquence used in its support, must be rejected as groundless.[33]

Students of polemics could learn much from this passage, from the damning with faint praise to the attempt to isolate the opposition from the scientific mainstream by casting aspersions of ignorance and hinting at ridicule.

Jameson goes on to argue as follows.

The Huttonians consider that gneiss and mica-schist are mechanical deposits because rocks formed from a state of complete fusion cannot retain a 'slaty' structure to any extent. If this inference is correct then such rocks could never occur in veins because according to the Huttonians all veins are filled by injection of fluid matter from below. It is well known, however, that all these rocks ('porphyry-slate', 'clay-slate', 'mica-slate', 'slaty quartz') *occur in veins which are often of immense magnitude.*[34]

Therefore they cannot be 'mechanical' in origin, formed from the alteration by heat of detrital sediments. Jamieson contrasts sandstone with gneiss and schist. In the former rock, feldspar, quartz, and mica occur as manifestly clatic particles bearing signs of attrition. In the latter, these minerals are clearly crystalline, with no cement between the grains.

It follows irresistibly from these facts, that gneiss and mica-slate are true chemical productions; and as gneiss passes into granite, and mica-slate into clay-slate, all these rocks are to be considered as belonging to the same species of formation. We have thus succeeded in proving, that the four great primitive rocks, Granite, Gneiss, Mica-slate, and Clay-slate, are not composed of materials derived from rocks older than themselves, but are in the strictest sense Primitive. Therefore the present world has not derived its materials from one or more worlds antecedent to it, and consequently this grand principle of the Huttonian Theory falls to the ground.[35]

Wavy strata or folds are attributed to irregular veins. The Huttonian theory of veins is criticized, with quotations from Playfair being cited at length, and Playfair's claim that veins are wider below than above is challenged on the basis of mining experience. Those veins that close at the top are not principal veins but only *branches* of them. It is shown[36] that some veins are filled with rocks such as mica-slate which according to the Huttonians never attained a condition of complete fusion!

Quite evidently Jameson's conception of veins was very different from that of the Huttonians, and embraced rock strata or bands of metamorphic rock. A satisfactory dialogue is always difficult to conduct between people with totally different preconceptions.

Elsewhere in his book Jameson flatly denies heat contact phenomena at the margins of whinstones, and denies that the substances produced by Hall in his experiments can be confused with rocks occurring naturally in the earth's crust.

In 1808 Jameson founded the Wernerian Society and edited its Memoirs from 1811. In the first two volumes the geological papers are all uncritically Wernerian, using the master's terminology, but by about 1820 the Memoirs had substantially abandoned this rigorous neptunism, although there was no dramatic recantation or change in outlook. Progressively more papers were published in zoology, botany, and palaeontology as the spirit of neptunism waned. Jamieson, to his credit (or perhaps he was simply short of material to publish), accepted articles both supportive of and antagonistic to the Wernerian doctrine. Generally, however, the far less numerous articles in support of plutonist views were published in the *Transactions of the Geological Society* with more popular articles for the layman appearing in the *Edinburgh Review*.

The dogmatism and *a priori* reasoning of the Wernerians was attacked in 1811 in withering terms by an anonymous critic believed to be W. H. Fitton, one of Jameson's most gifted students who subsequently defected to the Huttonians.

The Wernerian school obstructs the progress of discovery. The manner in which it does so is plain. By supposing the order already fixed and determined when it is really not, further enquiry is prevented, and propositions are taken for granted on the strength of a theoretical principle, that require to be ascertained by actual observation... When a Wernerian geognost, at present, enters on the examination of a country, he is chiefly employed in placing the phenomena he observes in the situations which his master has assigned to them in his plan of the mineral kingdom. It is not so much to describe the strata as they are, and to compare them with rocks of the same character in other countries, as to decide whether they belong to this or that series of depositions, supposed once to have taken place over the whole earth...[37]

Fitton recounts in a later article[38] an interesting anecdote concerning the Revd William Richardson, a staunch neptunist who supported Kirwan in his interpretation of the Portrush, 'basalt'. Visiting Edinburgh from his native Ireland, he was shown by Hall at Salisbury Crags a contact of sandstone and basalt with a displaced fragment of sandstone in the basalt, suggesting that the latter had intruded the sandstone. Richardson expressed contemptuous surprise that a theory of the earth should be founded on such trivial appearances!

The controversy was always more passionate in Edinburgh than in London and in fact a prime goal of the founders of the Geological Society in 1807 was to eschew argument and speculation in favour of sober fact-finding. Leading members of the society in the second decade of the century, such as Conybeare and Buckland, not only accepted basalts as volcanic but tended to support the Huttonians in their views on granite and the importance of heat in transforming rocks.

It must have been a bitter blow for Jameson when another of his star pupils defected to the Huttonian Camp. Ami Boué (1794–1881) was born in Hamburg of Franco-Swiss parents and studied medicine at Edinburgh but, like Hutton, his real passion was geology. In 1820 he wrote an essay exhibiting at least partial conversion to a plutonic interpretation of Scottish granites and in 1823 described granite intrusion in the Pyrenees. Although he acknowledged Hutton's priority, many continental geologists treated Boué as the original discoverer.[39] In 1822 he calmly asserted in the *Memoirs of the Wernerian Society* that the Erzgebirge, one of Werner's own stamping grounds, contained evidence that might be adduced in support of the Huttonian theory.

Adam Sedgwick was later to comment: 'For a long while I was troubled with water on the brain, but light and heat have completely dissipated it;[40] I did not examine them [certain rocks for fossils] in 1823–1824 because I thought that they were all below the region of animal life. At that time I had not quite learned to shake off the Wernerian nonsense I had been taught.'[41]

Jameson was stubbornly still teaching 'Wernerian nonsense' as late as 1825, to the dismay of at least one of his young students, Charles Darwin.

I heard the Professor at Salisbury Crags discoursing on a trap dyke with amygdaloidal margins and the strata indurated on either side, with volcanic rocks all around us, say that it was a fissure filled with sediments from above, adding with a sneer that there were men who maintained that it had been injected from beneath in a molten condition. When I think of this lecture, I do not wonder that I determined never to attend to Geology.[42]

Fortunately for geology, Adam Sedgwick and Charles Lyell aroused his enthusiasm a few years later.

As Jameson was not an unintelligent man and lived to a ripe old age, well beyond the time when the controversy had become mere history, it is worth inquiring whether he ever recanted. No such recantation ever appeared in print but according to Sweet[43] he is said to have done so at a meeting of the Royal Society of Edinburgh, but the precise date is not known.

Conclusions

From the time of Lyell onwards, historians of geology have tended to see the neptunist-vulcanist/plutonist controversy in a rather one-sided way. While conceding Werner's contribution to mineral and rock classification (which is after all a mere minor art) and qualities as a teacher, he is widely believed to have set back the progress of geology by his dogmatic insistence, regardless of evidence, on a wildly erroneous doctrine.[44] As for many of Werner's disciples, the contempt in which they were held by their opponents seems to have persisted for an unduly long time after the controversy was over. In Great Britain, if not so much on the continent, there was a strong association of neptunism with the more extreme catastrophist ideas, with rigid adherence to the Mosaic account of history in Genesis. While this is true of Kirwan and to a lesser extent de Luc, as will be noted in the next chapter, it is not true of many more distinguished geologists who supported neptunist doctrines, and decidedly not true of the master himself. Werner was in fact, like Hutton, a deist and was even accused of atheism.[45]

Desmarest, Hutton, and their supporters have been, in contrast, treated reverently as paragons of the empirical, inductive approach generalizing cautiously from carefully observed and soundly interpreted facts.

Rather than uncritically accepting the traditionally popular view, and seeing the eventual general acceptance of the Huttonian interpretation simplistically in terms of a triumph for the forces of light (or heat?) over the powers of darkness, or at least obscurantism, we should be provoked into wondering what it was about Werner's teaching that stimulated some of the best brains in Europe both into actively transmitting it to others and investigating geology over an extensive part of the globe. What indeed was so seductive about neptunist doctrine to generate so many crusaders? Before attempting to answer this question it is desirable to redress the balance somewhat between neptunist and plutonist interpretations of particular geological phenomena.

While it became almost universally accepted by the mid-1920s that rocks such as basalt and granite were not aqueous precipitates but igneous in origin, the neptunist interpretation that sediments became consolidated by compaction and mineral cementation is closer to the truth than Hutton's insistence on heat as the causal agent, even for the formation of flints in chalk. Even Playfair attributed the origin of a siliceous conglomerate to a stream of melted flint injected into a mass of loose gravel! Indeed, for many years one of the principal arguments against the Huttonians was that no amount of heat would on its own cause the consolidation of clastic sediments.

Again the Huttonians argued that *all* veins were igneous in origin, and had been fed from below. The neptunist alternative that they were low-temperature aqueous precipitates derived from above is now known to be at least partly correct for many metalliferous veins.

A further point to note is that, although uplift of the sea bed provides a better explanation of the occurrence of horizontal or marine strata in hills and mountains than subsidence of a universal ocean, Hutton was much vaguer about that part of his dynamic cycle which involved uplift than that involving denudation and sedimentation.

While it is easy for the present generation of geologists to be scornful about the failure of the neptunists to correctly interpret field evidence that to us appears straightforward and obvious, we have the benefits of an enormously greater knowledge of the natural world. It is worth recalling that, despite his pioneer work in the Auvergne, as good an observer as Guettard did not consider that basalt was a volcanic rock, while even Desmarest confessed that, given only experience of the basalts of Saxony, he would have made the same error as Werner. Even Buch, having convinced himself of the volcanic origin of basalts in the Auvergne, was for a long time reluctant to extrapolate to Saxony.

We must remember that study of rocks under the microscope did not commence until many years later, and the primitive knowledge of petrochemistry of the time is clearly indicated by Desmarest's belief that basalt could be derived from the melting of granite. Werner thought he could find convincing transitions between basalts and sediments, as did Jamieson between granite and gneiss, and both cited this evidence in support of the neptunist interpretation. The technique of tracing transitions is standard geological field practice and has always been one of the best means of establishing relationships.

Without doubt Werner can be criticized for generalizing so boldly to the earth from an extremely limited field acquaintance with rocks, and indeed his system began to crumble as soon as the most able of his students, such as d'Aubuisson, von Buch, and Humboldt, began to explore relevant rocks further afield. We must recognize, however,

that, while Desmarest appears to have been a straightforward empiricist, Hutton was as much a rationalist system-builder as Werner, as
many of the key field observations which provided critical evidence in
support were made long after his theory was first formulated. If he
was basically correct in ascribing greater significance to heat than
water in accounting for a wide range of geological phenomena, our
present vastly improved knowledge suggests he was often right for the
wrong reasons.[46]

There is one important respect in which Werner's system was
superior to Hutton's; it had a stratigraphy. Admittedly it was only a
crude lithostratigraphy but nevertheless a necessary beginning to a
proper historical study of the earth, which after all was recognized
from early times to be the essence of geology.[47] The local divisions of
Füchsel, Lehman, and Bergmann were useful advances, but only a
universal scheme underpinned, unlike those of the cosmogenists, by
hard factual data, could provoke intellectual excitement. It was the
idea of universal correlation which was surely what sent the Wernerians on their way across the world recording the presence of, for
instance, Primitive, Transitional and Floetz formations. In this
respect Fitton's strictures in his 1811 article on the Wernerians seem a
little unfair, for even today much perfectly respectable geological
research consists of converting the chaos of rock outcrops observed in
the field into maps and sections recording stratal dispositions, related
to a pre-established stratigraphic scheme.

Hutton's work, in contrast, has no stratigraphy whatsoever, and
hence no real history; it is instead a model of a system in dynamic
equilibrium, with Newtonian overtones of a temporally indeterminate
interplay of forces. In this, very fundamental, respect, Hutton's
system is not a rival to Werner's, and did not supersede it.[48] In fact
Hutton's *Theory of the earth*, monumental classic though it undoubtedly is, has a decidedly antique flavour for the modern reader
and not just because of the language it employs.

In my opinion the principal reason why the neptunist-plutonist
controversy faded so quickly in the early nineteenth century is
because of the development of a marvellous new research technique,
the correlation of strata by fossils. The importance of the pioneer work
of Cuvier and Brongniart (who acknowledged their debt to Werner)
in the Paris Basin, and more particularly of William Smith in
England, cannot be underestimated. More than anything else, this
work laid the basis for modern geology by allowing the establishment
of a valid relative time-scale which could be used across the world.

Neither the Wernerians nor the Huttonians had much interest in
fossils, and certainly give no hint of appreciating their stratigraphic

significance. Within about a couple of decades of active research stimulated by the great pioneering discoveries, the Wernerian strati-graphic scheme had been completely undermined. Rocks of 'Primitive' type in the Alps, in strata highly inclined to the horizontal and strongly metamorphosed, were found to contain fossils indicating that they were younger than unmetamorphosed, horizontally-disposed strata elsewhere. Werner's 'universal' formations turned out to be anything but universal.

Application of Smith's principles to the complex and obscure 'Transition' rocks in Britain led Sedgwick and Murchison to begin to unravel their secrets.

More generally, the whole pattern of geological research activity changed rapidly, and solutions were sought to newly emergent problems. Much of the research endeavour of the older generation began to seem rather irrelevant to the interests of the newer. One of the principal consequences of the widespread recognition of the biological and historical significance of fossils was the emergence of a new controversy whose significance for geology was so profound and wide-ranging that its echoes persist right up to the present. To this we now turn.

Notes

1. R. Porter (1977). *The making of geology: earth science in Britain 1660–1815*. Cambridge University Press.
2. For example: 'This surely cannot be considered in any other light than as a dream, formed upon a poetic fiction of a golden age' (J. Hutton (1795) *Theory of the earth*, Vol. 1, 271). '.. even Milton had scarcely ventured in his poem to indulge his imagination so freely ... as this writer, who set forth pretensions to profound philosophy' (C. Lyell (1830) *Principles of Geology*, Vol. 1, 37).
3. S. J. Gould (1887). *Time's arrow, time's cycle*. Havard University Press.
4. Ospovat, A. M. (1971), *Abraham Gottlob Werner, short classification and description of the various rocks*. Hafner Press, New York. The principal near-contemporary biographic account is the celebrated *éloge* by Cuvier (1819–27). In *Recueil des éloges historiques lus dans les séances publiques de l'Institut Royal de France*. Levrault, Strasbourg. There are good, briefer accounts of Werner's life and work in what are widely regarded as the three standard texts of the early history of geology: Geikie, A. (1897). *The founders of geology*. Macmillan, London; von Zittel, K. A. (1901). *History of geology and palaeontology to the end of the nineteenth century* (trans. M. M. Ogilvie-Gordon). Scott, London; Adams, F. D. (1938). *The birth and development of the geological sciences*. Baillière, Tindall, & Cox, London.

5. *Éloge* (translated by F. D. Adams), *op. cit.* (Note 4), p. 214.
6. Jameson, R. (1976), *The Wernerian theory of the Neptunian origin or rocks*. Hafner Press, New York. Facsimile reprint of *Elements of Geognosy* (1808).
7. See Ospovat, *op. cit.* (Note 4).
8. Hedberg, H. D. (1969). *Stockholm Contr. Geol.* **20**, **19**.
9. The idea of a universal ocean was an old one, dating back at least to Leibnitz and Steno.
10. Werner, A. G. (1971). *New theory of the formation of veins*; p. 10 of the English translation by C. Anderson (1809). Constable, Edinburgh.
11. J. Pinkerton (1811), *Petrology: a treatise on rocks*, Vol. 2. White, Cochrane, London.
12. See von Zittel, *op. cit.* (Note 4).
13. A good account, with full bibliographies, of the works of Guettard and Desmarest is given by Geikie, *op. cit.* (Note 4). For Desmarest see also Taylor K. L. (1969). In *Towards a history of geology* (ed. C. J. Schneer), p. 339. MIT Press.
14. Translation by Geikie, *op. cit.* (Note 4), P. 76.
15. von Zittell, *op. cit.* (Note 4) gives the best account of his continental work.
16. Werner, A. G. (1788). In *Neue Entdeckung, Intelligenz-Blattes des Allgem. Literatur Zeitung*, no. 57.
17. de Dolomieu, G. (1789). *Journal des Mines* **41**, 385.
18. Most of our contemporary information on Hutton is based on the biographic account by his friend and disciple John Playfair (1803). *Trans. Roy. Soc. Edin.* **4**, 39. E. B. Bailey (1967) gives a commentary on his writings in his book *James Hutton – the founder of modern geology*. Elsevier, Amsterdam, and R. H. Dott, Jr. (1969) gives a modern assessment of his scientific significance, in *Towards a history of geology* (ed. C. J. Schneer), p. 122. MIT Press.
19. Kirwan, R. (1794). *Trans Roy. Irish Acad.*
20. *Hutton, J. (1795). Theory of the earth, with proofs and illustrations.* 2 vols., Edinburgh. Facsimile reprint in 1959 by Wheldon and Wesley, Codicote, Herts.
21. Bailey (*op. cit.*, Note 18) suggests a possible excuse that Hutton was very ill for a time during preparation of the manuscript.
22. se Saussure, H. B. (1779–96). *Voyages dans les Alpes*. 4 vols. Barde-Magnet, Geneva.
23. The principal reference sources for this section are the books by Geikie and Adams cited in Note 4.
24. d'Aubuisson de Voisins, J. F. (1804). *J. Physique* **58**, 427; **59**, 367.
25. d'Aubuisson de Voisins, J. F. (1819). *Traité de Geognosie*, Vol. 2, p. 603. Translation by Geikie.
26. Buch, L. von (1802–9). *Geognostiche Beobachtungen auf Reisen durch Deutschland und Italien.* 2 vols. Hande & Spener, Berlin.
27. Buch, *op. cit.* (Note 26), Vol. 1, p. 126. Translation by Geikie.
28. It appears that his revised opinions were known to d'Aubuisson's referees when they recommended him to visit the Auvergne.

29. Buch, L. von (1810). *Reise durch Norwegen und Lappland*. Berlin. An English translation by J. Black was published by Colburn, London in 1813.
30. Playfair, J. (1802). *Illustrations of the Huttonian theory*. Edinburgh. Facsimile reprint in 1956 by University of Illinois Press.
31. Hall, Sir J. (1798). *Trans. Roy. Soc. Edin.* **5**, 43.
32. Jameson, R. (1808). *Elements of geognosy*. Facsimile reprint (1976) under the title *The Wernerian theory of the Neptunian origin of rocks*. Hafner Press, New York.
33. Jameson *op. cit.* (Note 32), pp. 344–5.
34. My italics.
35. Jameson, *op. cit.* (Note 32), p. 348.
36. Jameson, *op. cit.* (Note 32), p. 346.
37. Anon. (1811). *Edinburgh Rev.* 18, 95.
38. Fitton, W. H. (1837). *Edinburgh Rev.* 65, 9.
39. See Fitton, *op. cit.* (Note 38).
40. Clarke, J. W. and Hughes, T. McK. (1890). *Life and letters of Adam Sedgwick*, Vol. 1, p. 284. Cambridge University Press.
41. Clarke and Hughes, *op. cit.* (Note 40), p. 251.
42. Darwin, F. (1887). *Life and letters of Charles Darwin*, Vol. 1, p. 41. Murray, London.
43. Sweet, J. M. (1976). Introduction to facsimile reprint of Jameson, *op. cit.* (Note 32).
44. See, for example, Geikie, *op. cit.* (Note 4), p. 103. '...although he did great service by the precision of his lithological characters and by his insistence on the doctrine of geological succession, yet that as regard geological theory whether directly by his own teaching, or indirectly by the labours of his pupils and followers, much of his influence was disastrous to the *higher interests of geology*' my italics.
45. Ospovat, *op. cit.* (Note 4).
46. G. L. Davies (*The earth in decay: a history of British geomorphology, 1578–1878*. Macdonald, New York, 1969) was probably the first to challenge the empiricist myth promoted by Geikie, pointing out that Hutton developed his theory in final form before he had seen an unconformity, and when he had observed granite in only one inconclusive outcrop. Contemporary continental geologists put Hutton among the 'armchair' system builders of an earlier age. Porter, *op. cit.* (Note 1), in a refreshing counterblast to conventional Huttonian hagiography, notes how independent Hutton was of the empiricist mainstream of late-eighteenth century geologists in Britain, and argues that he can only be properly understood as a member of the Scottish Enlightenment along with other leading figures such as David Hume and Adam Smith.
47. Lyell's *Principles of Geology* starts with the following statement: 'Geology is the science which investigates the successive changes that have taken place in the organic and inorganic kingdoms of nature; it enquires into the causes of these changes, and the influence which they have exerted in modifying the surface and external structure of our planet.'

48. von Engelhardt, W. (1982) (*Fortschritte der Mineralogie* **60**, 21) argues that neptunism and plutonism represent the historic roots of two separate branches of geoscience. The former, which predates Werner, developed into historical geology, with its component disciplines of stratigraphy, palaeontology and palaeogeography, while plutonism marks the origin of the geodynamic approach concerned with the understanding of process, which treats of the underlying physico-chemical basis of natural phenomena.

2
Catastrophists and uniformitarians

Like the neptunists, with whom they have often been associated, the catastrophists have generally had a bad press. In the popular mythology they were the ones who gave free play to fantasy, who rashly invoked supernatural causes and allowed their geological researches to be dictated by *a priori* metaphysical beliefs. The uniformitarians on the other hand were sober-minded, sensible people who, guided by the talismanic principle that 'the present is the key to the past, eventually carried all before them and routed the opposition by their devotion to careful study of natural phenomena and cautious, empirically well-based and logically sound inference. The true history of the controversy is a good deal more complicated and certainly more interesting.

As Hooykaas[1] has pointed out, there has been a widespread confusion about the difference between *method* (i.e. research technique) and *system* (i.e. an all-embracing theory). *Catastrophism* is a system dating back to the speculative cosmogenies of Burnet, Whiston, and Woodward at the turn of the seventeenth and eighteenth centuries, which is no doubt a major reason why catastrophists of a later period were held up to ridicule. *Uniformitarianism* on the other hand, as outlined by Lyell, is both a system and a method. The English term uniformitarianism has frequently been used as an exact equivalent of the continental term *actualism*, and refers to the study of present-day processes as a means of interpreting past events.

Now it is perfectly possible to use actualistic methods and come to 'catastrophist' conclusions. Thus it was widely believed in the latter part of the eighteenth century, when the earlier cosmogenists were generally derided and a powerful empiricist ethos reigned, that the most dramatic features of the landscape, high mountain ranges with strangely contorted rocks at their core, and the deep gorges and valleys that cut into them, could not be accounted for in terms of present-day processes. The invocation of immensely powerful forces at some time in the past, unmatched in the modern world, seemed inescapable to many who would doubtless have been offended by an imputation that they sought the aid of the Almighty. Modern streams and rivers apparently contained too little water and did not flow fast enough to cut the valleys in which they flowed. It was far from obvious how the comparatively modest recent uplifts of land recorded, for instance, in the volcanic regions of Italy could have anything to do with the strongly tilted or contorted strata of the Alps and other mountain ranges. For those outstanding researchers who did not adopt neptunist beliefs, such as Dolomieu and Pallas, exceptionally violent change in former times seemed a quite logical inference.

If Hutton deserves the epithet 'founder of modern geology'[2] it is less for his plutonism than for challenging head-on the widespread contemporary belief in such a fundamental decoupling of past and present, and for introducing the notion of immense, if not indeterminate, time.

The Huttonian theory and its critics

Actualistic methodology had been used sporadically by geological researchers in the eighteenth century, even by the cosmogenist Buffon. The fact that the brilliant Russian scientist M. V. Lomonosov (1711–65), who died when Hutton was still a farmer and Werner a youth, employed actualistic principles[3] is probably significant for the following reason. Even with only restricted, second-hand contact with the intellectual centres of western Europe, a man of ability free of the mental shackles of rigid adherence to biblical doctrines could independently arrive at an approach that would later become commonplace.

Only Hutton, however, sought to use the actualistic method as a means of establishing some measure, however qualitative, of the passage of time. This is clearly indicated in the following extract from his 1788 paper,[4] which belies his reputation for obscurity.

In examining things present, we have data from which to reason with regard to what has been; and, from what has actually been, we have data for concluding with regard to that which is to happen here after. Therefore, *upon the supposition that the operations of nature are equable and steady*,[5] we find, in natural appearances, a means of concluding a certain portion of time to have necessarily elapsed, in the production of those events of which we see the effects.

Note here the prime assumption expressed in the italicized phrase. It seems to have the same logical status as the alternative, based on the striking appearance for instance of Alpine valleys, that the operations of nature are *not* equable and steady.

Later on Hutton goes further and introduced the notion of *indefinite* time.

To sum up the argument, we are certain,[6] that the coasts of the present continents are wasted by the sea, and constantly wearing away the whole; but this operation is so extremely slow that we cannot find a measure of the quantity in order to form an estimate. Therefore, the present continents of the earth, which we consider as in a state of perfection, would, in the natural operations of the globe, require a time indefinite for their destruction.[7]

Rigorous application of this seminal principle leads to a conclusion eminently satisfactory to a deist who is not concerned with first causes and who believes in a Newtonian mechanistic universe.

Therefore, there is no occasion for having recourse to any unnatural supposition of evil, to any destructive accident in nature, or to the agency of any preternatural [supernatural] cause, in explaining that which actually appears.[8]

Nowhere in his paper, incidentally, does Hutton cite any recent scientist who had actually invoked the supernatural, though no doubt popular myths persisted.

Though the above passages hardly suggest that paragon of empiricism of geological folk-lore, the Huttonian principle had great heuristic value for dynamical geology, when compared with any alternative. Consider for instance the eloquent passage in Playfair's *Illustrations* where he discusses the vexed question of mountain valleys.

It is, however, where rivers issue through narrow defiles among mountains, that the identity of the strata on both sides is most easily recognised, and remarked at the same time with the greatest wonder ... there is no man, however little addicted to geological speculations, who does not immediately acknowledge, that the mountain was once continued quite across the space in which the river now flows; and if he ventures to reason concerning the cause of so wonderful a change, he ascribes it so some great convulsion of nature, which has torn the mountain asunder, and opened a passage for the waters. It is only the philosopher [scientist], who has deeply meditated on the effects which action long continued is able to produce, and on the simplicity of the means which nature employs in all her operations, who sees in this nothing but the gradual work of a stream, that once flowed as high as the top of the ridge which it now so deeply intersects, and has cut its course through the rock, in the same way, and almost with the same instrument, by which the lapidary divides a block of marble or granite.[9]

By employing what I consider to be one of the most beautiful arguments in the whole of his book Playfair continues his attack on the catastrophist interpretation of valleys.

Every river appears to consist of a main trunk, fed from a variety of branches, each running in a valley proportional to its size, and all of them together forming a system of valleys, communicating with one another, and having such a nice adjustment of their declivities, that none of them join the principal valley, either on too high or too low a level, a circumstance which would be infinitely improbable if each of these valleys were not the work of the stream that flows in it.

If, indeed, a river consisted of a single stream without branches, running in a straight valley, it might be supposed that some great concussion, or some powerful torrent, had opened at once the channel by which its waters are conducted to the ocean; but, when the usual form of a river is considered, the trunk divided into many branches, which rise at a great distance from one another, and these again subdivided into an infinity of smaller ramifications, it becomes strongly impressed upon the mind that all these channels have been cut by the waters themselves; that they have been slowly dug out by the washing and erosion of the land; and that it is by the repeated touches of the same instrument that this curious assemblage of lines has been engraved so deeply on the surface of the globe.

A scientific theory is worthless if it fails to stimulate the imagination. By his own account, Playfair's imagination as regards the immensity of geological time was deeply stimulated in 1788 when he visited, in company with Hutton and Hall, the famous unconformity at Siccar Point, where subhorizontal Devonian sandstones rest on near-vertical Silurian slates and greywackes.

Dr. Hutton was highly pleased with appearances that set in so clear a light the different formations of the parts which compose the exterior crust of the earth, and where all the circumstances were combined that could render the observation satisfactory and precise. On us who saw these phenomena for the first time, the impression made will not easily be forgotten. The palpable evidence presented to us, of one of the most extraordinary and important facts in the natural history of the earth, gave reality and substance to those theoretical speculations, which, however probable, had never till now been directly authenticated by the testimony of the senses. We often said to ourselves, what clearer evidence could be had of the different formation of these rocks, and of the long interval which separated their formation, had we actually seen them emerging from the bottom of the deep? We felt ourselves necessarily carried back to the time when the schistus on which we stood was yet at the bottom of the sea, and when the sandstone before us was only beginning to be deposited, in the shape of sand or mud, from the waters of a superincumbent ocean. An epocha still more remote presented itself, when even the most ancient of these rocks, instead of standing upright in vertical beds, lay in horizontal planes at the bottom of the sea, and was not yet disturbed by that immeasurable force which has burst asunder the solid pavement of the globe. Revolutions still more remote appeared in the distance of this extraordinary perspective. The mind seemed to grow giddy by looking so far into the abyss of time; and while we listened with earnestness and admiration to the philosopher who was unfolding to us the order and series of these wonderful events, we became sensible of how much further reason may sometimes go than imagination may venture to follow.[10]

Let us note how, in this celebrated passage, Playfair rather lets his imagination run away with him in extrapolating back through time. 'Revolutions still more remote appeared in the distance of this extraordinary perspective. The mind seemed to grow giddy by looking into the abyss of time.' The hard evidence was restricted to only one Huttonian 'cycle'. Furthermore, he implies in this and the previously cited passages that the immense time periods that need to be invoked relate more to the denudational than the uplift part of the cycles, '... when even the most ancient of these rocks, instead of standing upright in vertical beds, lay in horizontal planes at the bottom of the sea, and was not yet *disturbed by that immeasurable force which has burst asunder the solid pavement of the globe*'. The words I have italicized have a strong connotation of catastrophic uplift, indicating a mode of thinking quite conventional at the time. This decided presumption of asymmetry in the cycles seems to reflect accurately the thoughts of the master himself. Thus Hutton refers to '... that enormous force of which regular strata have been broken and displaced; ... strata, which have been formed in a regular manner at the bottom of the sea, have been violently bent, broken and removed from their original place and situation.'[11]

As a mathematician Playfair permits himself a calculus metaphor concerning his attitude to geological time.

It affords no presumption against the reality of this progress, that, in respect of man, it is too slow to be immediately perceived. The utmost portion of it to which our experience can extend, is evanescent, in comparison to the whole, and must be regarded as the momentary increment of a vast progression, circumscribed by no other limits than the duration of the world. TIME performs the office of *integrating* the infinitesimal parts of which this progression is made up; it collects them into one sum, and produces from them an amount greater than any than can be assigned.[12]

Hutton concluded his 1788 paper with his most famous sentence. 'The result, therefore, of our present enquiry is, that we find no vestige of a beginning - no prospect of an end.' Playfair expands this in the following way.

The Author of nature has not given laws to the universe, which, like the institutions of men, carry in themselves the instruments of their own destruction. He has not permitted, in his works, any symptoms of infancy or of old age, or any sign by which we may estimate either their future of their past duration.[13]

Hutton's theory was such a radical departure from anything that has gone before that an adverse critical reaction was virtually inevitable.[14] Even his friend Sir James Hall was no easy convert.

I was induced to reject his system entirely, and should probably have continued to do so, with the great majority of the world, but for my habits of intimacy with the author, the vivacity and perspicuity of whose conversation formed a striking contrast to the obscurity of his writings. I was induced by that charm, and by the numerous original facts which his system had led them to observe, to listen to his arguments in favour of opinions which I then looked upon as visionary. After three years of almost daily warfare with Dr. Hutton on the subject of his theory, I began to view his fundamental principles with less and less repugnance.[15]

As has already been pointed out in Chapter 1, published work criticizing Hutton's 1788 paper came mainly from Kirwan and De Luc.

Richard Kirwan (1733–1812) was President of the Royal Irish Academy from 1799 to 1819 and a chemist and mineralogist of some repute. In his 1793 article[16] he attacked Hutton in the most vehement terms, ridiculing and misrepresenting him and accusing him of atheism. Kirwan thought it absurd that it should be a philosophical necessity that we must remain ignorant of the origin of the earth. According to him it is easy to reconcile the neptunist succession of rocks with the Mosaic account; geological exegesis unfolds the Genesis story step by step. Though no geologist, Kirwan claimed a wide acquaintance with natural phenomena but this is manifestly not so. His main objection to Hutton is not based on evidence but on his failure to support a literal belief in Genesis. In view of the transparent special pleading in Kirwan's paper, and the expanded version of his case present in his *Geological essays* (1799) it is surprising that his attack attracted such serious attention, although he made a valuable point in arguing for the *historicity* of geology, which Hutton had overlooked. In retrospect its greatest value seems to be that it spurred Hutton into providing full documentation for his theory.

In his reply to Kirwan in Chapter 2 of Volume 1 of *Theory of the earth*, Hutton contemptuously dismisses Kirwan's views on granite. Kirwan's conclusion, in reaction to Hutton's 'no vestige of a beginning', was 'Then this system of successive worlds must have been eternal', but Hutton replied 'Such is the logic by which, I suppose, I am to be accused of atheism'. He goes on to claim that Kirwan has read into his words meaning he had not intended.

In tracing back the natural operations which have succeeded each other, and mark to us the course of time past, we come to a period in which we cannot see any further. This, however, is not the beginning of those operations which proceed in time . . . My principal anxiety, was to show how the constitution of this world has been widely contrived; and this I endeavoured to do, not from supposition or conjecture, but from it answering so effectively the end of its intention, viz. the preserving of animal life, which we cannot doubt being its purpose.

In other words Hutton shies back from adopting the Aristotelian view of an eternal world, pleading a methodological restriction; the geological evidence is inadequate to decide and he is content to leave it at that. Nevertheless there is no hint of progression or directionality in his system.

Jean André de Luc (1727–1817) was an altogether more substantial critic. He was born and grew up in Geneva but lived in England from 1773. An accomplished naturalist as well as being well versed in both physics and chemistry, he was considered to be one of the pioneers in the study of geological dynamics. He was perhaps best known among contemporaries for his championing of the widely accepted view that there is a fundamental difference between causes now in action and those which have ceased to act (such as forces to elevate mountains). Thus it was inevitable that he would come into conflict with Hutton.

De Luc published four letters in response to Hutton's 1788 paper[17] and later summarized his views in an elementary treatise on geology.[18] The crux of his system was that earth history is subdivisible into two intrinsically dissimilar eras, the earlier one involving the formative processes that give rise to continents, the later one involving events since that time. The earlier era, which had ceased operations, was subdivisible into six successive stages corresponding to the Wernerian formations. It was brought to an end approximately 4000 years ago by a striking event which marked the commencement of the modern era. In a great catastrophe ancient lands were inundated by a huge mass of rushing water.

All this is held as if to be independent of the Mosaic account and de Luc was delighted to find a close correspondence between scientific evidence and Genesis, especially the Deluge.[19] In contrast to his scornful dismissal of Kirwan, who in his *Geological essays* had committed the unforgivable sin of reasoning *a priori* from Genesis itself, de Luc was respectful towards Hutton and Playfair, against whose notions, however, his treatise was primarily directed. He was especially worried about the great antiquity of the world that was demanded.

According to the conclusion of Dr. Hutton, and of many other geologists,[20] our continents are of indefinite antiquity, they have been peopled we know not how, and mankind are wholly unacquainted with their origin. According to my conclusions, drawn from the same source, that of *facts*, our continents are of such small antiquity, that the memory of the revolution which gave them birth must still be preserved among men; and thus we are led to seek in the book of Genesis the record of the history of the human race from its origin. Can any object of importance superior to this be found throughout the circle of natural science?[21]

It was a common feature of scientific debating practice at the time to accuse opponents of wilfully forcing facts into line with subjective delusions, and de Luc so accused Hutton and Playfair. As the quotation indicates, he was convinced that facts were on his side but that, of course, depends on what one means by facts. The 'facts' that persuaded Playfair that shorelines had risen indicated to de Luc that sea level had fallen!

Cuvier's *Discours*

In the early years of the nineteenth century Edinburgh was rivalled and eventually eclipsed by Paris as a centre of geological research and foundations were laid there for the science of palaeontology. This was due essentially to the founding by the new republic of the Natural History Museum in the Jardin des Plantes and the appointment there as professors of several men of exceptional talent. Geology was represented by Faujas de Saint-Fond, mineralogy successively by Haüy and Brongniart, invertebrates by Lamarck, and vertebrates by Geoffrey Saint-Hilaire and, the most brilliant of all, Cuvier.

Though generally thought of as French, Georges Cuvier (1769–1832)[22] had Swiss parents and completed his formal education in Stuttgart. His great work was on the comparative anatomy and functional morphology of fossil vertebrates, which contained the establishment of the 'law of correlation of parts', and the sensational confirmation, from study of fossil elephants and giant sloths, that some organisms had become extinct. Cuvier also made a substantial advance in stratigraphy. Influenced by the classic work in Saxony and Thuringia of Lehman and Füchsel, Cuvier and Brongniart turned their attention to the strata of the Paris Basin, paying full attention, unlike earlier workers, to the fossil content. A whole series of faunas was recognized back through the Tertiary to the Cretaceous Chalk, which appeared to end (in a time sequence) as suddenly as they began. Using actualistic methods Cuvier and Brongniart were able to demonstrate that the Tertiary strata represented cyclic alternations of freshwater and marine conditions, which appeared to provide a possible confirmation of Hutton's ideas. (Another great Frenchman, Desmarest, had expressed scepticism about Huttonian cycles on grounds of insufficient supporting evidence.)

Cuvier expanded and generalized his ideas in a major publication which appeared in 1812, a year after the Paris Basin memoir was published. This was devoted primarily to his researches on fossil bones, but he introduced in the preface a new theory as a *Discours*

Préliminaire. This preface was quickly recognized to be of major importance and went through several editions as *Discours sur les révolutions de la surface du globe*. Jameson published an English translation which went through four editions.[23]

Cuvier considered himself an empiricist *par excellence* who refused to indulge in speculations that could not be firmly backed up by evidence.

We now propose to examine those changes which still take place on our globe, investigating the causes which continue to operate on its surface . . . This portion of the history of the earth is so much the more important, *as it has long been considered possible to explain the more ancient revolutions on its surface by means of these still existing causes*[24] . . . But we shall presently see that unfortunately this is not the case in physical history; *the thread of operations is here broken,*[24] the march of nature is changed, and none of the agents that she now employs were sufficient for the production of her ancient works.[25]

It is clear from this passage that actualism was nothing new to Cuvier, and indeed it had been practised on the continent by, among others, Pallas, Saussure, and Dolomieu. Nevertheless Cuvier firmly allies himself with de Luc in inferring a decoupling of Recent and Ancient. It is important to note that Cuvier, like de Luc, considers that his system had strong empirical support.

The essential elements of Cuvier's ideas are contained in the following passage.

The lands once laid dry have been reinundated several times, whether by invasions of the sea or by transient floods; and it is further apparent to whoever studies the regions liberated by the water in its last retreat, that these areas, now inhabited by men and land animals, had already been above the surface at least once—possibly several times—and that they had formerly sustained quadrupeds, birds, plants, and terrestrial productions of all types. The sea, therefore, has now departed from lands which it had previously invaded. The changes in the height of the waters did not consist simply of a more or less gradual and universal retreat. There were successive uprisings and withdrawals of which, however, the final result was a general subsidence of sea level.

But it is also extremely important to notice that these repeated inroads and retreats were by no means gradual. On the contrary, the majority of the cataclysms that produced them were sudden. This is particularly easy to demonstrate for the last one which by a double movement first engulfed and then exposed present continents, or at least a great part of the ground which forms them. It also left in northern countries the bodies of great quadrupeds, encased in ice and preserved with their skin, hair and flesh down to our times. If they had not been frozen as soon as killed, putrefaction would have decomposed the carcasses. And, on the other hand, this continual frost did not previously occupy the places where the animals were seized by the ice, for they could not have existed in such a temperature. The animals were killed, therefore, at the same instant when glacial conditions overwhelmed the countries they inhabited.

This development was sudden, not gradual, and what is so clearly demonstrable for the last catastrophe is not less true of those which preceded it. The dislocations, shiftings, and overturnings of the older strata leave no doubt that sudden and violent causes produced the formations we observe, and similarly the violence of the movements which the seas went through is still attested by the accumulations of debris and of rounded pebbles which in many places lie between solid beds of rock.

Life in those times was often disturbed by these frightful events. Numberless living things were victims of such catastrophes: some, inhabitants of the dry land, were engulfed in deluges; others, living in the heart of the sea, were left stranded when the ocean floor was suddenly raised up again; and whole races were destroyed forever, leaving only a few relics which the naturalist can scarcely recognise.[26]

The evidence for the episodic catastrophes included the sudden disappearances of faunas up the stratigraphic succession, for which his personal experience was restricted to the Paris Basin, alternations of freshwater and marine strata in the Paris Basin, and conglomerates and stratal disturbance in older rocks. His inference of sudden death among Siberian mammoths, remains of which had been found in permafrost terrains in the late-eighteenth century, was given a heavy weighting in support.[27] We may note the highly confident, even dogmatic, tone of pronouncement allied with limited field experience and recall the much-respected Werner.

Though officially Christian, like virtually all his scientific contemporaries, Cuvier showed no published interest in attempting to reconcile the geological record with Genesis, let alone to invoke supernatural causes for his catastrophes; the principal focus of his attention was to determine the record of past events as precisely as he could. While the successive catastrophes, whose origins were not speculated upon, caused the extinction of animal species, the species that replaced them in the stratal succession were thought possibly to have migrated from other continents or seas. The implication is that not all the catastrophes, though extensive, had necessarily been world-wide. Cuvier did not speculate on how the new species originated, but he strongly opposed the evolutionary views of his colleague Lamarck.

The establishment of catastrophist doctrine

Cuvier's collaborator, Alexandre Brongniart, was able to demonstrate in 1821 the presence of Cretaceous fossils at about 2000 metres altitude in the Savoy Alps and fossils resembling those of the Paris Basin Tertiary high in the Vicentine Alps.[28] This work clearly indicated that fossils are the key to correlation, and that some mountains are young in age. This type of research was pursued further by Cuvier's most brilliant disciple in France, Leonce Elie de Beaumont (1798–1874), whose work also had great influence in Britain and Germany. Utilizing the new stratigraphic technique established by Cuvier and Brongniart in France and Smith in England, he made a detailed study of folded strata in mountain ranges over a wide area in Europe, and was able to prove that mountain-building episodes had occurred at many different times. Of particular interest was his recognition that a major episode of mountain uplift in the Pyrenees occurred between the Cretaceous and Tertiary, the time of a mass faunal extinction. The strong implication was accepted that extinctions could be caused by the elevation of mountain ranges if they were sufficiently sudden.[28]

Elie de Beaumont followed Cuvier in arguing that folded and tilted strata implied sudden disturbance, and that one was not entitled to extrapolate to such 'catastrophic' phenomena from the manifestly slow and gradual 'causes now in operation'. Long periods of quiescence were interrupted suddenly by relatively short-lasting upheavals of the land or torrential inundations of the sea.

Another important French contribution came from Adolphe Brongniart, the son of Alexandre, who in 1823 published a monograph on fossil floras from the Upper Palaezoic to the Recent.[29] Not only the animals but the plants exhibited changes through successive formations. He thought that the Carboniferous coal measure floras of western Europe had a tropical character, which provided biological confirmation of a contemporary geophysical theory, based on the geothermal gradient recorded in mines. The mathematical physicist Fourier considered that the geothermal gradient was due to the earth's residual heat, with the rate of cooling gradually diminishing through time. Elie de Beaumont incorporated this notion into his mountain-building synthesis. The earth had shrunk as it cooled, causing the compression of strata into folds and the uplift of mountain ranges. Likewise, vulcanicity was more intense in former times when the earth was hotter.

The intellectual power of this synthesis as it appeared to well-informed contemporaries should not be underestimated. Significant elements of neptunism, involving progression from older, highly deformed, to younger, undeformed rocks, and plutonism, involving igneous activity and mountain uplift, were brought together, with the implication of greater expenditure of geotectonic energy in earlier times. Major phases of extinction and replacement in the organic world could be related to these tectonic changes involving sudden uplift and subsidence. The new stratigraphic tool provided by fossils could be used to correlate past events over wide areas and prove a whole succession of mountain-building episodes by the recognition and dating of angular unconformities. As Adam Sedgwick put it, referring to the 'breakthrough' pioneered by Cuvier,

The beautiful conclusions drawn from unexpected facts; the happy combination of mineralogical and zoological evidence; the proofs of successive revolutions, till then unheard of in the physical history of the earth – all these things together, not merely threw new light on a subject before involved in comparative darkness, but gave new powers and new means of induction to those who should in after times attempt any similar investigation.[30]

In Great Britain, catastrophist theory had become well established, along with Huttonian dynamics and plutonism, in the 1820s. A good idea of contemporary thought is revealed by an influential elementary textbook by one of the leading geologists of the time, the Revd W. D. Conybeare,[31] The crude formulations of the neptunists were upgraded into the new stratigraphy based on correlation by fossils. Traprocks were volcanic in origin and strata had originally been deposited horizontally. Uplift and tilting of strata was thought of as catastrophic, as clearly indicated by the choice of words in Conybeare's introduction. '[Geologists] will undoubtedly attribute to the *explosions* which produce them, the principal agency in forcing up the strata, and *heaving* from the depths of the waves the ponderous masses of the continents.'[32]

Conybeare, along with virtually all his scientific contemporaries, was a firm believer in the scriptural Deluge, and Jameson, in his editorial notes to the translated version of Cuvier's *Discours*, endeavoured to equate Cuvier's most recent revolution with this. (Preoccupation with the Deluge was of much more concern in Britain than on the continent, perhaps partly because some of the leading figures, such as Buckland, Conybeare, and Sedgwick, were clergyman.)[33]

Without question the leading diluvialist (or student of the Deluge) was William Buckland (1784–1856), who in 1813 was appointed Reader in Mineralogy in the University of Oxford and in 1818 became Reader in Geology. Though geology was not an examination requirement his open lectures were of such excellence that a large number of distinguished men, including heads of Oxford colleges, were attracted to them, and his pupils included both Lyell and, outside the university context, also Murchison. A much loved and revered eccentric,[34] his lecture room at the Old Ashmolean Museum and, subsequently, in the Clarendon Building, was a jumble of rocks and bones. He laid great emphasis on fieldwork, and always appeared on field excursions dressed in his top hat and with his blue bag to hold rocks and fossils.

In his inaugural lecture[35] as Reader of Geology, Buckland endeavoured to demonstrate that the facts of geology are consistent with the accounts of creation and the Deluge recorded in the Mosaic writings.

In all these [geological and palaeontological phenomena] we find such undeniable proofs of a nicely balanced adaption of means to ends, of wise foresight and benevolent intention and infinite power, that he must be blind indeed, who refuses to recognise in them proofs of the most exalted attributes of the Creator.[36]

William Paley, the most celebrated exponent of the argument for design, could hardly have expressed it better. Buckland considered that the six days of creation were to be taken figuratively, and long periods of time might have elapsed between successive acts of creation. The earlier, literal interpretation of Genesis as exemplified by Kirwan was to be discarded; prior to the Deluge the earth could have considerable antiquity. As for the Deluge itself, the geological evidence for it was incontrovertible.

Buckland completed his empirical theology *before* he stumbled across the evidence which won it the most widespread support.[37] Quarrymen had in 1821 unearthed a rich fauna of fossil bones in Kirkdale Cavern, north Yorkshire. Buckland was led to study the bones, which included the remains of hyenas with subordinate lions, elephants, rhinoceroses, hippopotamuses, etc., and inferred the former presence of a den of hyenas.[38] This work led on to a more extensive study, with colleagues, of many cave deposits in Europe, utilizing his thorough knowledge of comparative anatomy which had been inspired by Cuvier. The publication of his *magnun opus* entitled 'Relics of the flood'[39] won him paeans of critical praise.

According to Buckland, the Deluge had been *universal*, and therefore could not precisely correspond to Cuvier's most recent revolution which had been attributed to a general interchange in position of continents and oceans. Most bones in the European cave deposits consist of the remains of extinct species, which are never found in river alluvium. As there is no human testimony to their post-flood existence the bones must have been interred in the caves prior to the time of Noah. The top layers of bones are so perfectly preserved in sediments that they must have been buried suddenly. Judging by the amount of post-diluvial stalagmite-covered mud, this event could not have taken place more than about 5000 years ago. That such relics turned up only in caves illustrates the violence of the catastrophe that engulfed them. The uniform mud deposits indicated that only one flood has occurred since. Buckland was profoundly disturbed not to find human remains that could be referred to ante-diluvial times. To Buckland the end of the hyena epoch was produced by Noah's deluge. *Because* no good evidence of ante-diluvian humans was found with the hyena assemblage, he later antedated the Deluge to *before* Man's creation.

Buckland was widely read, and used reports of fossil bones found at high altitude in the Andes and Himalayas as evidence that the great diluvial event was not confined to low ground but was deep enough to cover the higher mountains. He mustered a wide range of evidence in support of the Deluge. This included: gorges and ravines cut through mountains, buttes, and mesas and immense deposits of gravel, together with scattered boulders on hills or slopes where no river could have carried them. It was impossible to refer such phenomena to the puny modern agents of erosion and sediment transport. Buckland in fact supported Sir James Hall's interpretation of some kind of huge torrent resembling a tidal wave of enormous magnitude.

The most illustrious of Buckland's supporters, the Woodwardian Professor of Geology at Cambridge, Adam Sedgwick (1785–1873) stressed the scientific character of the investigations into the Deluge.

It must ... at once be rash and unphilosophical [i.e. unscientific] to look to the language of revelation for any direct proof of the truths of physical science. But truth must at all times be consistent with itself. The conclusions established on the authority of the sacred writings may, therefore, consistently with the soundest philosophy, be compared with the conclusions established on the evidence of observation and experiment ... the application is obvious. The sacred record tells us – that a few thousand years ago 'the foundations of the great deep were broken up – and that the earth's surface was submerged by the water of a general deluge; and the investigations of geology tend to prove that the accumulations of alluvial matter have not been going on many thousand years; and that they were preceded by a great catastrophe which has left traces of its operation in the diluvial deritus which is spread out over all the strata of the world.

Between these conclusions, derived from sources entirely independent of each other, there is, therefore, a general coincidence which it is impossible to overlook, and the importance of which it would be most unreasonable to deny. The coincidence has not been assumed hypothetically but has been proved legitimately, by an immense number of direct observations conducted with indefatigable labour, and all tending to the establishment of the same general truth.[40]

The uniformitarian challenge

It required another scholarly clergyman, the Revd John Fleming, a member of the Wernerian Society, to call into question the closeness of correspondence between the Mosaic testimony and the geological and palaeontological evidence. Moses left word of a gentle strand of water rising placidly for 40 days, with the flood leaving no trace except a rainbow. This was hardly the account of a violent and transient torrent. Further, Cuvier insisted on a *complete* extinction of species whereas according to Moses two individuals of each were saved. Fleming thought it unscientific to extrapolate so boldly beyond the evidence as Buckland had, in for instance attributing random Andean and Himalayan records of bones to the deluge. 'Diluvial' and alluvial sediments in different parts of the world had been too readily grouped together as deposits of the same event, which was quite unproven. Furthermore, there was ample evidence for the *gradual* excavation of valleys. The extinction of the 'diluvial' fauna, like the excavation of valleys, was not sudden but piecemeal and occupied a long time period. Mixtures of extinct and extant species were known from some bone-bearing gravels.[41]

George Poulett Scrope (1797–1876), who later became a Member of Parliament, was another important critic of diluvial theory in the mid-1820s. He thought that the attempt to reconcile geological research with scripture threatened the integrity and status of geology and deplored its anti-heuristic character: 'it stops further enquiry'. He became well known for a book on volcanoes[42] and more particularly for a classic monograph on the volcanic terrain of the Massif Central,[43] making important advances on the work of Desmarest and others.

The rambling subtitle of Scrope's general book on volcanoes included *theory of the earth*. Scrope's theory was orthodox in that it was directionalist and progressive, with diminution of energy through time as the earth cooled, and new organisms appearing in succession. It was also catastrophist, as the following passage indicates.

These sudden partial elevations of the crust of the globe, and the other various causes which at this period disturbed the tranquillity of the primitive ocean, produced violent waves and currents, which broke up and triturated the projecting eminences of its bottom, and distributed their fragments in alluvial conglomerate strata, wherever the turbulence of these moving waters was partially checked.[44]

Scrope's originality lay in his postulating that the geological agents he invoked were all still in operation, though with diminished energy.

There were indeed occasional 'excesses of disturbance' (e.g. to elevate the Alps) but, unlike Cuvier, he did not believe that the thread of operations had been broken. Scrope emphasized the presence of volcanic rocks in strata of all epochs, and stressed the impossibility of a simple twofold classification into antediluvian[45] and post-diluvian periods.

He showed by detailed description of the volcanic terrains in Central France how volcanoes of many different ages could be classified according to the degree to which they and their lava flows had been eroded by running water. Some valleys showed alternating beds of lava and river gravels. Volcanic activity almost entirely post-dated freshwater formations. Long intervals of volcanic quiescence were implied.

Playing a Huttonian theme and demanding 'almost unlimited drafts upon antiquity' he argued that there was no need to invoke volcanic activity in Central France of greater activity than elsewhere at the present day. The present valley of the Loire and its tributaries within the Le Puy Basin have been excavated since lava was erupted, sections though which occur on the valley sides. Yet the lavas are undeniably contemporary because cones of loose scoriae occur locally. These would have quickly been removed by any violent rush of water. It follows that the erosive forces of the existing streams must be sufficient to excavate the valleys. Furthermore, it was difficult to see how a rushing torrent could have excavated the complex bends in flood-plain rivers known as meanders. Scrope thought that one could extrapolate such results to other mountainous regions where the evidence of volcanic rocks is lacking.

Not a great deal of attention seems to have been paid to Scrope's work by what we might call the orthodox catastrophists; they were readily prepared to grant large amounts of time for erosion provided it were antediluvian. In his Presidential Address to the Geological Society in 1830 Sedgwick was readily prepared to concede with Scrope, that river meanders argue against catastrophism, and referred to other evidence favouring gentle rates of river erosion. But Scrope's ideas were, he thought, conspicuously unsuccessful in accounting either for far-travelled erratic boulders and the gravels of the Thames Valley and elsewhere. Processes operating at the present day seemed too feeble to account for these phenomena.

The work of Fleming and, more especially, Scrope was to have a great influence on the man who was to become the high priest of a new doctrine, uniformitarianism.

Though he spent most of his life in England, Charles Lyell (1797–1875)[46] belonged by birth to the Scottish landed gentry. Born at

Kinnordy, Angus, his interest in geology was aroused at Oxford by the lectures of Buckland. Though he trained for a career in law he practised it without enthusiasm for only two years, using his weak eyesight as a convenient excuse to pursue full-time what had become his passion. Only for two years, 1831–3, however, despite the eminence he achieved, did he hold a professional position in geology, as Professor at Kings College, London. Subsequently he depended for his income on private means and book royalties. (*Principles of geology* went through 11 editions, the last appearing in 1875.)

In 1822 he examined Cretaceous strata in the Weald under the tutelage of Gideon Mantell, who had recently discovered reptile bones and plant remains in what appeared to be the sand and clay deposits of a freshwater delta. Lyell was struck by the fact that these deposits underlay the marine deposits of the Chalk, after deposition of which there had evidently been significant uplift and erosion. Here was very clear indication, on London's 'doorstep', of the erroneousness of Wernerian doctrine.

In the following year Lyell paid a lengthy trip to Paris where he met, among others, Cuvier, Brongniart, and Humboldt. However, the French geologist who influenced him most was Constant Prévost, and he travelled extensively with him on several occasions. Prévost's researches in the Paris Basin had demonstrated how a succession of small changes could convert a sea into a freshwater lake. Prévost was a pioneer in identifying and recognizing the significance of lateral variations of facies at the same stratigraphic horizon, and made a persuasive case for the Paris Basin in Tertiary times as an inlet of the sea, with the salinity diminishing eastwards as the distance from the open marine connection increased. This work raised serious doubts about the drastic character of environmental change claimed by Cuvier, but Prévost's attempts to challenge the great man's authority in France met with little success.

Lyell paid a visit to the family estate in Angus in 1824 and spent some time in the vicinity studying freshwater marls in the beds of some small lakes. These were found to contain disintegrated specimens of the green alga *Chara*. Conversion of these calcareous marls into a rock would result in the production of a limestone closely comparable to if not identical with the freshwater limestone in the Paris Basin that Cuvier and Brongniart thought had no modern counterpart.

In an article published in 1826 Lyell followed orthodox opinion in referring to strata having been subjected at various times to violent convulsions. The influence of Scrope is apparent, however, in the following passage.

But in the present state of our knowledge, it appears premature to assume that existing agents could not in the lapse of ages, produce such effects as fall principally under the examination of the geologist. It is an assumption, moreover, directly calculated to repress the ardour of inquiry by destroying all hope of interpreting what is obscure in the past by an accurate investigation of the present phenomena of nature.[47]

Scrope's memoir on the Massif Central provoked a strong desire in Lyell to visit the region, which he did in 1828 in the company of Roderick Murchison.

They soon saw evidence of the power of rivers to cut deep gorges through lava beds. The lavas had filled old valleys and rivers had then cut new channels. Lyell was impressed by the discovery in the Limagne that certain Tertiary *Chara*-bearing limestones closely resembled deposits forming today near his Scottish home. Indeed, freshwater formations were very extensive. One particularly striking limestone contained the fossilized cases of caddis fly larvae, which are common in modern lakes and streams. The limestone lithology reminded him, however, of English Jurassic examples, emphasizing to him the importance of using fossils for dating purposes and stressing the limited validity of the old Wernerian lithostratigraphy.

Lyell and Murchison also found marl strata, exceeding 215 metres in thickness, which were usually composed of laminae, about 15 per centimetre. Each lamina consisted of the plate-like valves of the ostracod *Cypris*. Lyell knew *Cypris* from his collecting experience in pools and ditches in Hampshire, and concluded that the layers were annual. Therefore the whole formation must have represented several hundred thousand years of deposition in calm water in conditions resembling those of certain lakes at the present day.

It appeared that the freshwater strata were elevated most strongly where they directly overlaid the huge volcanic centre. This fact seems to have first suggested to Lyell an intimate causal connection between vulcanism and tectonic disturbance. If the spectacular uplift and folding strata in mountain regions could be ascribed simply to long-continued action of ordinary earthquakes, there would be no grounds for postulating occasional paroxysmal convulsions. Hence a keystone would be removed from the catastrophic interpretation.

The full implications of his observations in the Massif Central appear to have dawned on Lyell only while he was waiting in Nice for Murchison to recover from illness. He then decided to go to Sicily on Murchison's recommendation and was there to make some discoveries crucial to his whole planning and conception of the *Principles*.[48]

At Padua, Lyell and Murchison parted, the latter to return to England. Lyell continued to the Naples region, there to climb

Vesuvius and visit the so-called 'temple' of Serapis, which had undergone an interesting history of subsidence and uplift since Roman times. On the volcanic island of Ischia he was intrigued to discover strata containing extant species of Mediterranean molluscs at an elevation of about 1840 metres. If Ischia had risen this much in the geologically recent past, how much more elevation might be anticipated for Etna?

In Sicily he first approached the greatest volcano in Europe by way of the Val de Bove, a huge natural amphitheatre some eight kilometres across, floored by recent lava. Buckland had told him that this was an especially worthwhile place to visit, perhaps because he thought that it supported Buch's craters-of-elevation hypothesis, which implied that the whole mountain might have been uplifted in one paroxysmal event, but probably rather more because it was a very good geological site. Lyell found sections, however, in the surrounding cliffs revealing alternating layers of lava and tuff dipping uniformly outward from the centre of the mountain. This seemed to imply a gradual and therefore prolonged period of growth. He also discovered the shells of marine molluscs interbedded with lava flows a few hundred metres above sea level, suggesting recent uplift.

Lyell was deeply impressed by the phenomena he encountered when he climbed to the summit of Etna. Out of some 80 parasitic cones on its huge flanks, only one, Monte Rossi, had erupted in the last few centuries. If even a quarter of this number formed in the 30 centuries since human records began, then a period of at least 12 000 years must have elapsed to account for all 80. Yet these were only the most recent cones not yet buried by lava flows.

Further exciting discoveries were made in the Val di Noto to the south, where some low hills are composed mainly of massive limestone containing moulds of marine shells. These were underlain around Syracuse by a blue clay with much better preserved shells that could be demonstrated to be identical with those of extant species in the Mediterranean. So the limestone with a Mesozoic-like appearance must also be extremely young Tertiary. All the strata, though evidently so young in geological terms, apparently extended under and hence predated Etna. Yet the argument from the cones indicated that by a human time-scale the volcano was of immense antiquity. The limestone reached as high as 100 metres in the interior of Sicily, but evidently there had been time enough for the uplift to have been gradual not paroxysmal.

On his return to Rome from Sicily in January 1829, Lyell wrote to Murchison about his plans for a new book which would deal with the principles of reasoning in geology. These principles 'are neither more nor less than that no causes whatever have from the earliest time to which we can look back, to the present, ever acted but those now acting and that they never acted with different degrees of energy from that which they now exert.'[49]

Volume 1 of the book appeared in 1830 and Volumes 2 and 3 were published within three years. Its title indicates that it is no mere textbook but a treatise devoted to the presentation and defence of a new system: *Principles of geology, being an attempt to explain the former changes of the earth's surface by reference to causes now in operation.* Volume 1 opens with the earliest comprehensive account of the history of geology from the time of the Ancients. As Rudwick[50] has pointed out, Lyell's historical account is designed to show that the science had been retarded in recent years firstly by attempts to reconcile geology with scripture and secondly by the assumption that past causes must have been different from those of the present. Indeed, he is somewhat indiscriminate in his attacks on the Mosaic and Catastrophist approaches to geology.

Lyell felt that the authority of the popular 'progressionist' view had to be undermined at the outset, and stressed the inadequate conception of the geological time-scale. He needed to alter the scientific *imaginations* of his fellow geologists who were prepared intellectually to admit the vastness of geological time.

Lyell's methodology is firmly rooted in an actualism that demands an assumption of the constancy of physical laws.

Our estimate, indeed, of the value of all geological evidence, and the interest derived from the investigation of the earth's history, must depend entirely on the degree of confidence which we feel in regard to the permanency of the laws of nature. Their immutable constancy alone can enable us to reason from analogy, by the strict rules of induction, respecting the events of former ages, or, by a comparison of the state of things at two distinct geological epochs, to arrive at the knowledge of general principles in the economy of our terrestrial system.[51]

This belief in the heuristic value of the actualistic method stems directly from Hutton, as adopted successively by Playfair and Scrope, but in promoting a steady state, uniformitarian 'system' Lyell goes further than Hutton.

There can be no doubt, that periods of disturbance and repose have followed each other in succession in every region of the globe, but it may be equally true, that the energy of the subterranean movements have been always uniform as regards the whole earth. The force of earthquakes may for a cycle of years have been invariably confined, as it is now, to large but determinate spaces, and may then have gradually shifted its position, so that another region, which had for ages been at rest, became in its turn the grand theatre of action.[52]

Now Hutton had written: '... we are not to limit nature with the uniformity of an equable progression ...',[53] hardly a uniformitarian sentiment. He also envisaged catastrophic uplift, whereas Lyell saw no need to invoke such drastic effects. In his view, erosion is balanced by deposition, subsidence by elevation, in a system of steady state with fluctuations about a mean. Lyell shared Hutton's belief in a relationship between earthquake activity, vulcanicity, and uplift, and in his rejection of any notion of progression or directionality through time.[54]

In his rejection of progressionism, Lyell attacked the conception of a cooling earth. There was no evidence that igneous and tectonic activity had been more intense in former times, or that the world climate had cooled systematically. Laudan[55] emphasizes the use Lyell made of the *vera causa* method, or method of true causes, as promoted by one of his intellectual mentors, the Cambridge scholar John Herschel, and illustrates this with reference to his theory of climatic change. According to Lyell, changing climates could be explained by changing geography, which could oscillate with time rather than show a directional tendency. Now geography was *observable* whereas the earth's central heat was not. If land accumulated around one of the poles, for example, extensive cooling would ensue, whereas if it accumulated close to the equator there would instead be warming. Ultimately, however, Lyell's climatic theory depended on his elevation theory, which was poorly developed, with no *vera causa* principle being applied.

The intense vulcanicity of Italy, the Massif Central, and elsewhere was very recent, and was preceded by long periods of quiescence. Older rocks contained more granitic and metamorphic[56] materials only because they had formed at depth and were only exposed after significant uplift and erosion; this was more likely the older the rocks. In the Bernese Alps it had recently been demonstrated that strata as young as Tertiary had been metamorphosed into schists.[57]

Lyell repeatedly returned to the phenomenon of uplift, because as we have seen his notion of slow and gradual movement was one of his most original conceptions; it struck at the heart of Elie de Beaumont's and von Buch's popular theories. A time analogy is drawn (in Volume 1) with the building of the pyramids of ancient Egypt. We should only be warranted in ascribing the erection of the great pyramid to superhuman power if we were convinced that it was raised in a day.

Again and again Lyell stressed the ampleness of geological time to account for what can be taken superficially as evidence of catastrophic events. Thus the sharp faunal contrast and lithological break between the Cretaceous and Eocene is the result of an unrecorded time span perhaps longer than that separating the Eocene from the present.

The *Principles* culminates in an account of Tertiary stratigraphy, a subject to which Lyell made a lasting contribution. He had learned that, whereas Sicilian beds contained mostly extant species, the Subapennine beds of Italy had many extinct species and the Tertiary of Nice an even greater proportion. In turn, the Tertiary of the Paris Basin had significantly fewer living species than Nice. He had realized therefore by the end of his 1828–9 tour of France and Italy that the Tertiary must consist of several periods long enough to allow accumulation of thousands of feet of sediments and to allow significant organic change. Over the next few years, in collaboration with the great Parisian conchologist Gérard Paul Deshayes, Lyell developed a classification of the Tertiary into Eocene, Miocene, older Pliocene, and younger Pliocene, based on the proportion of extinct species.[58]

When Lyell surveyed the evidence of volcanic activity in successive Tertiary periods he saw that as volcanic activity subsided in one area it started elsewhere. There were many shifts and long quiescent intervals. The notion accepted by some that Tertiary volcanic activity was more intense than today was probably due to the age of the Tertiary being grossly underestimated.

It might be thought the Lyell's full awareness of the facts of organic change, as recorded in the Tertiary strata of which he had direct experience, would have persuaded him to concede at least some progression through time. On the contrary he argued that the evidence of change provided by fossils is illusory because of the selective non-preservation in older strata of higher organisms such as mammals. Organisms would, moreover, come and go with changing climate. Envisaging for the sake of his argument a future with more benign climate than the present day, he stated: 'Then might those genera of animals return, of which the huge iguanodon might reappear in the woods, and the ichthyosaur in the sea, while the pterodactyle might flit again through umbrageous groves of tree-ferns'.[59]

Lyell's argument seemed especially weak in the case of our own species. In this case the absence of relics in older strata is evidence of recent appearance. That this might appear to others to be a contradiction was glossed over on the grounds that the coming of Man was 'an era in the moral, not in the physical world'.

The critical response to Lyell

As the organization of *Principles of geology* illustrates, Lyell was extremely skilful at presenting a case; his training as a barrister had not been wasted. He was also an astute tactician. He was delighted to learn that Scrope was to review *Principles* in the *Quarterly Review*, a renowned organ of the orthodox establishment, and endeavoured to brief him in a series of letters. 'If we don't irritate, which I fear that we may . . . we shall carry all with us. If you don't triumph over them, but compliment the liberality and candour of the present age, the bishops and enlightened saints will join us in despising both the ancient and modern physio-theologians. It is just the time to strike, so rejoice that, sinner as you are, the *Quarterly Review* is open to you... If Murray [the publisher] has to push my volumes, and you wield the geology of the *Quarterly Review*, we shall be able in a short while to work an entire change in public opinion.'[60]

In the event it was in the *Quarterly Review* that one of Lyell's leading critics, William Whewell (1794–1866), Professor of Mineralogy at Cambridge, coined the terms catastrophism and uniformitarianism.[61] The term catastrophism fails however, to do justice to the cluster of beliefs that characterized the opposition, because *directionalism*[62] was also an important component of their system.

The controversy between catastrophists and uniformitarians, which enlivened the meetings of the Geological Society through the 1830s, never descended to the depths of acrimony that had characterized the neptunist-vulcanist/plutonist controversy. In fact there was a good deal of mutual respect and even friendship between the leading participants, and Lyell was widely praised both for his wide knowledge and the thoroughness with which he pursued his actualistic method.

On one subject serious debate soon ceased. The *Principles* contributed to the demise of the diluvial theory. Although Conybeare's articles[63] replying to Lyell are perhaps the fullest statement of the diluvialist counter-offensive he backtracked from his earlier belief in the existence of evidence for the Mosaic Deluge; similarly Buckland dropped the Deluge as a geological agent in his contribution to the series of volumes devoted to the celebration of natural theology known as the Bridgewater Treatises.

In his Presidential Address to the Geological Society Sedgwick was uncompromising in his apostasy.

Having been myself a believer, and, to the best of my power a propagator of what I now regard as philosophical heresy . . . I think it right, as one of my last acts before I quit this Chair, thus publicly to read my recantation.

We ought, indeed to have paused before we first adopted the diluvian theory, and referred all our old superficial gravel to the action of the Mosaic Flood. For of man, and the works of his hands, we have not yet found a single trace of the remnants of a former world entombed in these deposits.[64]

Sedgwick, Conybeare, and Whewell joined battle with Lyell on what to them were more substantial issues.

In the same presidential address, Sedgwick is for the most part generous in his compliments to Lyell, but criticizes him for his *a priori* reasoning and compares him to a skilful advocate. From the very title page he presents himself as the defender of a theory, thereby offending against the inductive method of science.

Sedgwick fully recognized the need to apply an actualistic method in interpreting geological phenomena.

For we all allow, that the primary laws of nature are immutable – and that we can only judge of effects which are past, by the effects we behold in progress.

But to assume that the secondary combinations arising out of the primary laws of matter, have been the same in all periods of the earth, is ... an unwarrantable hypothesis, with no *a priori* probability, and only to be maintained by an appeal to geological phenomena.[65]

Sedgwick was greatly impressed by the work of Elie de Beaumont, whose recognition of angular unconformities of different ages in the mountain ranges of Europe seemed to him to provide decisive evidence in favour of long periods of repose being interrupted suddenly by brief cataclysmic events.

Conybeare stressed the directionalist view, and addressed himself to the changes that took place through time as a consequence of the earth cooling. Volcanic activity indeed required a heat source deep beneath the crust. Older strata were usually much contorted because of the expenditure of greater amounts of geotectonic energy; the Tertiary was more tranquil than older periods. The Alps were elevated in a single convulsion.

No real philosopher, I conceive, ever doubted that the physical causes which have produced the geological phenomena were the same in kind, however they may have been modified as to the degree and intensity of their action, by the varying conditions under which they may have operated at different periods. It was to these *varying conditions* that the terms, a different order of things, and the like, were, I conceive, always intended to have been applied; though these terms may undoubtedly have been by some writers incautiously used.[66]

Whewell, in adopting a similar line of attack, is particularly penetrating.

Time, inexhaustible and ever accumulating his efficacy, can undoubtedly do much for the theorist in geology; but *Force*, whose limits we cannot measure, and whose nature we cannot fathom, is also a power never to be slighted: and to call in one, to protect us from the other, is equally presumptuous to whichever of the two our superstition leans.[67]

In reality when we speak of the uniformity of nature, are we not obliged to use the term in a very large sense, in order to make the doctrine at all tenable? It includes catastrophes and convulsions of a very extensive and intense kind; what is the limit to the violence which we must allow to these changes? In order to enable ourselves to represent geological causes as operating with uniform energy through all time, we must measure our time by long cycles, in which repose and violence alternate; how long must we extend this cycle of change, the repetition of which we express by the word *uniformity*?

And why must we suppose that all our experience, geological as well as historical, includes more than *one* such cycle? Why must we insist upon it, that man has been long enough an observer to obtain the *average* of forces which are changing through immeasurable time.[68]

Last but by no means least is the evidence for organic progression.

It is clear ... that to give even a theoretical consistency of his system, it will be requisite that Mr. Lyell should supply us with some mode by which we may pass from a world filled with one kind of animal forms, to another, in which they are equally abundant, without perhaps one species in common. He must find some means of conducting us from the plesiosaurs and pterodactyls of the age of the lias, to the creatures which mark the oolites or the iron-sand. He must show us how we may proceed from these, to the forms of those later times which geologists love to call by the sounding of names of the paleotherian and mastodontean periods. To frame even a hypothesis which will, with any plausibility, supply the defect in his speculations, is a harder task than that which Mr. Lyell has now executed. We conceive it undeniable (and Mr. Lyell would probably agree with us) that we see in the transition from an earth peopled by one set of animals, to the same earth swarming with entirely new forms of organic life, a distinct manifestation of creative power, transcending the known laws of nature: and, it appears to us, that geology has thus lighted a new lamp along the path of natural theology.[68]

Whewell could hardly have hinted more strongly at a belief in persistent divine intervention. Sedgwick ardently espoused similar views, which indeed remained popular among the devout well into the mid-century. To challenge such a view risked incurring the charge of materialism. When one bears in mind that clerical critics and religious fundamentalists such as John Keble[70] in Oxford continued to fulminate against even such conservative geologists as Buckland and Sedgwick, it is no wonder that Darwin procrastinated so long before publishing his theory of natural selection.

The extracts cited above might give the impression of a concerted onslaught directed at Lyell, but this would be misleading because the *Principles of geology* was generally received with great enthusiasm and soon began to exert an enormous influence as a geological compendium. As has been noted, even his strongest critics praised his application of actualistic methodology and more specifically his contributions to elucidating Tertiary stratigraphy. Nor were British geologists in the 1830s as strongly polarized in their opinions as in the earlier controversy between neptunists and their opponents. Thus a leading geologist such as Henry de la Beche, the first director of the Geological Survey, could write that 'The difference in the two theories is in reality not very great; the question being merely one of intensity of forces, so that probably, by uniting the two, we should approximate nearer the truth.[71]

Even such supporters of Lyell as Scrope and Prévost did not share his extreme uniformitarian views, accepting the notion of progression or directionality in so far as they believed in a slowly cooling earth, with intensity of forces diminishing through time, and accepted the fossil evidence of change at its face value. Indeed the only notable figures who appear to have accepted Lyell's doctrine with little or no qualification in the early years were Gideon Mantell, who wrote a text giving a popular account for laymen, and Charles Darwin (1809–82) for whom the first volume of Lyell's *Principles* was one of his most treasured possessions on the voyage of the *Beagle*. (Before the voyage he had taken a crash course in geology from Sedgwick.)

At Saint Jago in the Cape Verde Islands Darwin undertook his first geological fieldwork of the voyage, which led to an exercise in Lyellian reasoning, in which he argued for a gradual elevation of the whole island and for later gradual subsidence around volcanic craters. It was in Chile, however, that Darwin became a complete convert to Lyell's system. Thus he quickly inferred that the Plain of Coquimbo must have been elevated at least 85 metres in the fairly recent past because the fossil shells scattered on the plain showed a great resemblance to those that could be picked up on the present-day beach. More generally, he recognized a systematic relationship between degree of elevation and the ratio of extant to extinct species; the higher the ratio the lower the elevation. Especially astonishing was the discovery of Jurassic fossils many thousands of feet above sea level.

The earthquake Darwin experienced in February 1835 had a considerable effect on his thinking, for the ground was raised along the coast by several feet for as much as 100 miles. Earthquakes of such magnitude apparently took place with a frequency of about one per century. Quite clearly, repetition of several thousand such small-scale events could elevate a mountain range as high as the Andes in much less than a million years.

Darwin's evidence of gradual, and still-continuing, elevation of the Andes threw strong doubt on Elie de Beaumont's theory. Indeed, catastrophic uplift of the Andes had been specifically mentioned as a possible cause of a 'diluvial' tidal wave. In 1838 he presented a paper to the Geological Society giving a detailed account of the connection between earthquakes, volcanic activity, and elevation of land in South America;[72] this was an explicit application of the Lyellian doctrine. Lyell was present at the meeting and, needless to say, welcomed the paper with enthusiasm.

By general consent, however, Darwin's greatest geological success was his theory of coral reefs, founded on his observations in the Pacific and Indian oceans after the *Beagle* left South America. In the second volume of the *Principles* Lyell had argued that coral atolls are built on extinct volcanoes that rose to just beneath the surface of the ocean. Pointing out, among other things, the unlikelihood of so many volcanoes rising to just the same height, Darwin argued instead for gradual subsidence of volcanic islands accompanied by upward and outward growth of corals. Initially volcanic islands will be surrounded by fringing reefs, then by barrier reefs detached by sea from land, and finally atolls will appear, consisting of a ring of coral reef with a central shallow lagoon. All these reef types were in evidence in various parts of the Pacific, and Darwin was using what has long been recognized as a standard geological method of inference, arguing from comparison of present-day features to an evolutionary time sequence.

Although Darwin's interpretation differed from Lyell's, it was Lyellian in inspiration, combining a thorough-going actualism and gradualism with a steady-state model of the earth's behaviour, subsidence in some areas, such as the Pacific atolls, being balanced elsewhere by elevation, as in the Andes. Although Darwin's theory proved a popular one, it did not receive evidential support until many decades later, when deep boreholes penetrated a few thousand metres of coral rock overlying basalt.

As detailed stratigraphic researches continued a number of important discoveries were made which supported Lyell. The better the regional stratigraphy was known, the more local the evidence of stratal disturbance. This progressively undermined the notion of extremely extensive episodes of cataclysmic disruption. Furthermore, uplift of mountain ranges such as the Alps could not simply be reduced to a single 'convulsion'. If mountain uplift were so local in space, the credibility of Elie de Beaumont's idea of relating faunal discontinuities to tectonic disturbance was reduced. Extensive areas were discovered in eastern Europe and North America where strata as old as Early Palaeozoic were flat-lying and unmetamorphosed, whereas many contorted Alpine schists were found to be Mesozoic or even Tertiary in age. The more that was known of the relevant stratigraphy, the less likely it seemed that vulcanicity had declined in intensity through the Tertiary. These various facts tended to render less plausible the notion of tectonic and igneous forces diminishing through time as the earth cooled. No reputable geologist, furthermore, seriously disputed the immensity of geological time, a fact still disturbing to the religious fundamentalists.

On the other hand Murchison's work on the Silurian System revealed a fauna confined to marine invertebrates, with no marine or land vertebrates or vascular plants. Destruction of fossils by metamorphism could be ruled out, so it became increasingly difficult to deny some form of organic progression.

By the middle of the century those who adhered stubbornly to the old catastrophist doctrine, such as Elie de Beaumont, Buch, Agassiz, Sedgwick, and Murchison, had become increasingly isolated figures. Uniformitarianism was the dominant doctrine in Britain and, to a somewhat lesser extent, on the continent. Lyell could therefore look back with considerable satisfaction on two decades of research which seemed to settle the issue beyond reasonable doubt. He chose the occasion of the two addresses he was required to give as President of the Geological Society to present his current assessment in the light of the newly acquired knowledge.

He noted in the first address[73] that the major controversy about mountain ranges concerned the twin catastrophist beliefs that (a) geotectonic energy had declined through time and (b) that short intervals of violent convulsions had interrupted long periods of quiescence. He referred to a recent monograph by Murchison on the structure of the Alps, Apennines, and Carpathians. Locally in the Glarus Alps of Switzerland, ammonite-bearing Jurassic limestones had been found to be overlain by schists; elsewhere they rested on Eocene flysch; the whole region exhibited spectacular inversions of strata. Murchison's interpretation was that in former times the crust of the earth had been affected by forces of *infinitely greater intensity that those which now prevail* (Murchison's words). To this Lyell replied as follows.

It is not the magnitude of the effects, however gigantic their proportions, which can inform us in the slightest degree whether the operation was sudden or gradual, insensible or paroxysmal. It must be shown that a slow process could never in any series of ages give rise to the same results.[74]

Old sea beaches had recently been found in Norway at 230 metres above sea level, containing shells of molluscan species still living in the North Sea. Therefore uplift must have taken place a comparatively short time ago. How much more stupendous then must have been the geographic changes between formations with very different faunas, between for example the Cretaceous and Tertiary.

Geologists had been forced to admit that both the Alps and Pyrenees had been subjected to more than one episode of tectonism, and Elie de Beaumont had had to abandon his earlier ideas that the Pyrenees had experienced only one paroxysmal uplift in favour of six or seven chronologically distinct systems of dislocation. In the case of the Alps, uplift of the flysch had been succeeded by deposition of the molasse which had then in turn been uplifted. As for Elie de Beaumont's claim that granite intrusions had become rarer in more recent times, the 'evidence' was easily accounted for by postulating that there had been as yet insufficient time for the younger granites to be exposed by erosion. We remained very ignorant of geological activity deep below the surface of the earth.

The imagination may well recoil from the vain effort of conceiving a succession of years sufficiently vast to allow of the accomplishment of contortions and inversions of stratified masses like those of the higher Alps; but its powers are equally incapable of comprehending the time required for grinding down the pebbles of a conglomerate 8000 feet [2650 metres] in thickness. In this case, however, there is no mode of evading the obvious conclusion, since every pebble tells its own tale. Stupendous as is the aggregate result, there is no escape from the necessity of assuming a lapse of time sufficiently enormous to allow of so tedious an operation.[75]

We can see that Lyell had no new arguments to offer but was simply like a patient teacher, providing fresh illustrations of his basic doctrine. In a similar way his thoughts had undergone no advance on the subject of the sequence of fossils revealed in the stratigraphic record, despite the many new discoveries over the previous two decades. The whole of his second address[76] was devoted to this subject.

As regards the origin of species Lyell rejected the heretical notion of 'transmutation' recently put forward by Robert Chambers, thereby endorsing the refutations by Sedgwick, Richard Owen, and Hugh Miller.

By the creation of a species, I simply mean the beginning of a new series of organic phenomena, such as we usually understand by the term 'species'. Whether such commencements be brought about by the direct intervention of the First Cause, or by some unknown Second Cause of Law appointed by the Author of Nature, it is a point upon which I will not venture to offer a conjecture.

Though Sedgwick did not believe in 'transmutation' he accepted the idea of organic progression ('creative additions'), as did the brilliant German palaeontologist Heinrich Georg Bronn, from less to more complex and perfectly organized forms. Thus Sedgwick saw in the apparent succession of cephalopods, fish, reptiles, mammals, and Man something analogous to the Great Scale of Being beloved of eighteenth century savants. Adolphe Brongniart had recognized something comparable in fossil plants, with a sequence from algae successively to ferns and horsetails, conifers and cycads, and palms and oaks.

As something that struck at the heart of his strictly steady-state conception, Lyell would still have none of this. It was quite understandable that the oldest, Lower Palaeozoic, flora was relatively primitive because all the deposits so far discovered were laid down in the sea; more highly organized forms no doubt lived on the land. Furthermore, the contemporary marine invertebrate fossils were all highly developed examples of Cuvier's three great divisions of Radiata, Articulata, and Mollusca; evidently marine invertebrates were as perfect then as today. The scarcity of fish and absence of marine mammals was hardly surprising in the present state of knowledge, bearing in mind their rarity in modern dredgings of the sea bottom.

In the Upper Palaeozoic reptiles of advanced organization were known, as were highly organized trees such as conifers and even palms, and Triassic and Jurassic fish and reptiles were as advanced as today. Traces of both birds and flowering plants were known in rocks of Triassic and Jurassic age and the rarity or absence of Mesozoic mammals is a consequence of collection failure. Mammals were well diversified early in Tertiary time and there was no indication of their progression through this era.

One senses the tone of special pleading becoming somewhat more strident than two decades earlier. Man is still placed on a pedestal and in effect detached from the rest of creation, as Lyell's concluding remarks indicate.

In the first publication of the Huttonian theory, it was declared that we can neither see the beginning nor the end of that vast series of phenomena which it is our business as geologists to investigate. After sixty years of renewed enquiry, and after we have greatly enlarged the sphere of our knowledge, the same conclusion seems to me to hold true. But if any one should appeal to such results in support of the doctrine of an eternal succession, I may reply that the evidence has become more and more decisive in favour of the recent origin of our own species. The intellect of man and his spiritual and moral nature are the highest works of creative power known to us in the universe, and to have traced out the date of their commencement in past time, to have succeeded in referring so memorable an event to one out of a long succession of periods, each of enormous duration, is perhaps a more wonderful achievement of Science, than it would be to have simply discovered the dawn of vegetable or animal life, or the precise time when out of chaos, or out of nothing, a globe of inanimate matter was first formed.[77]

Achievement for the second time of the presidency of the Geological Society in middle age must have seemed to Lyell in many ways to have marked the pinnacle of his career, after which he could look back with benign satisfaction at the science he had transformed so significantly. Unfortunately he was to receive a double blow within only a few years. A formidable challenge was mounted, and was to persist until the end of the century, to his steady-state model by a physicist whose prestige in his own discipline matched that of Lyell in his. Even more disturbing, one of his favourite disciples was to argue for, with impressive supporting evidence, a theory of organic evolution that put paid once and for all to a strictly uniformitarian view of the fossil record. We must postpone a modern assessment of the status of the catastrophist-uniformitarian debate until account has been taken of these later developments.

Notes

1. Hooykaas, R. (1970). *Kon. Nederl. Ak. Wetensch. aft. Letterkunde*, Med. (n.r.) **33**, 271.
2. For example, Bailey, E. B. (1967). *James Hutton – the founder of modern geology*. Elsevier, Amsterdam. I cannot agree with Bailey that Hutton was *the* founder of modern geology, because of the ahistorical nature of his work, and geology is nothing if not an historical science.
3. Tikhomirov, V. V. (1969). In *Towards a history of geology*. (ed. C. J. Schneer), p. 357. MIT Press.
4. Hutton, J. (1788). *Trans. Roy. Soc. Edin.* **1**, 217.
5. My italics.
6. Note the Wernerian assurance.
7. Hutton, *op. cit.* (Note 4), p. 301.
8. Hutton, *op. cit.* (Note 4), p. 285.
9. Playfair, J. (1802), *Illustrations of the Huttonian theory*, pp. 86–7. Creech, Edinburgh.
10. Playfair, J. (1803). *Trans Roy. Soc. Edin.* **4**, 39.
11. Hutton, J. (1788). *Trans. Roy. Soc. Edin.* **1**, 267, 284.
12. Playfair, J. *op. cit.* (Note 9), p. 117.
13. Playfair, J. *op. cit.* (Note 9), p. 119. This passage seems to indicate that, despite playing down the providential tone of Hutton's writing, he shared his deistic beliefs.

14. It has been suggested that Hutton might have plagiarized the work of an obscure country doctor, G. H. Toulmin, who was a medical student in Edinburgh from 1776 to 1779. A persuasive case has been made out, however, for the reverse because Toulmin could easily have learned something of Hutton's ideas either from unpublished manuscripts or Joseph Black's lectures (Davies, G. L. (1967). *Bull. geol. Soc. Am.* **78**,121).

    More pertinently, Davies, G. L. (1964). *Proc. Geol. Ass. Lond.* **75**, 493, pointed out that 'by 1668 Hooke had formed the shadowy outline of a theory of the earth that is almost identical with the theory that James Hutton presented to the Royal Society of Edinburgh in 1785. Hooke deserves to be remembered as a precursor of Hutton.' Robert Hooke's ideas were published posthumously in 1705. Drake, E. T. (1981) *Am. J. Sci.* **281**, 963, has argued convincingly that Hutton was well aware of Hooke's writings, although he makes no acknowledgment of him by name.

    Nevertheless the reaction of the geological community to Hutton is such that he is rightly seen as the real innovator, for the same sorts of reasons that the theory of evolution is first and foremost associated with the name of Darwin, despite the fact that he had a number of precursors.

15. Hall, J. (1812). *Trans. Roy. Soc. Edin.* **6**, 71.
16. Kirwan, R. (1793). *Trans Roy. Irish Acad.*
17. de Luc, J. A. (1790–1). *Monthly Rev.* **2**, 206; 582; **3**, 573; **5**, 564.
18. de Luc, J. A. (1809). *An elementary treatise on geology.* Rivington, London.
19. An excellent account of late-eighteenth and early-nineteenth century geological researches in Great Britain relating to the Genesis story in general, and the Deluge in particular, is given by Gillispie, C. C. (1951). *Genesis and geology.* Harvard University Press.
20. Evidently by 1809 Hutton and Playfair had acquired a reasonable number of supporters.
21. de Luc, J. A. *op. cit.* (Note 18), p. 82.
22. 'Georges' was actually a pseudonym, his correct names being Leopold Chrétien Frederic Dagobert! For his services to science and to the republic he was eventually created a Baron. A good account of his huge contribution to palaeontology is given by Rudwick, M. J. S. (1972). *The meaning of fossils.* Macdonald, London.
23. Cuvier, G. (1817–27, 4 editions). *Essay on the theory of the earth.* Blackwood, Edinburgh. Note that Jameson changed the title in a way that could be held to suggest some affinity with the cosmogenists. Cuvier is unlikely to have approved of this, since he held the cosmogenists in as much scorn as his contemporaries.
24. My italics.
25. Cuvier, G. *op. cit.* (Note 23), 3rd edn., p. 12.
26. Cuvier, G. (1825) *Discours* (3rd edn.) Vol. 1, pp. 8–9. Translation by C. C. Gillispie, *op. cit.* (Note 19).

27. We would not accept his reasoning today, of course. Mammoths could have lived actively in summer months in permafrost terrains. Occasionally one might have been trapped in surface muds late in the season, shortly before they froze up for the winter at a rate quick enough to prevent total putrefaction.

28. de Beaumont, L. Elie, (1829–30). *Ann. Sci. Nat.* **18**, 5–25; 284–416; **19**, 5–99, 177–240.

29. See Rudwick, *op. cit.* (Note 22).

30. Sedgwick, A. (1830). *Proc. Geol. Soc.* **1**, 197.

31. Conybeare, W. D. and Phillips, W. (1822). *Outline of the geology of England and Wales*. Phillips, London.

32. Conybeare and Phillips, *op. cit.* (Note 31), p. xvii. My italics.

33. Gillispie, *op. cit.* (Note 19).

34. It was proud boast of Buckland that he had eaten his way through the animal kingdom. One anecdote (possibly apocryphal) has it that, on being reverently shown the heart of a French king, carefully preserved in a snuff box in a house near Oxford, he promptly picked it up and swallowed it!

    W. Tuckwell's (1900) *Reminiscences of Oxford*, Murray, London, contains the following absorbing passage.

    > I recall ... when I was wont to play with Frank Buckland and his brother [Buckland's children] in their home at the corner of Tom Quad [Christchurch College]: the entrance hall with its grinning monsters on the low staircase, of whose latent capacity to arise and fall upon me I never quite overcame my doubts; the side table in the dining room covered with fossils, 'paws off' in large letters on a protecting card; the very sideboard candlesticks perched on saurian vertebrae; the queer dishes garnishing the dinner table – horseflesh I remembered more than once, crocodile another day, mice baked in batter on a third day – while the guinea pig under the table inquiringly nibbled at your infantine toes, the bear walked round your chair and rasped your hand with file-like tongue, the jackal's fiendish yell close by came through the open window...
    >
    > At Palermo, on his wedding tour, (Buckland) visited St. Rosalia's shrine ... It was opened by the priests, and the relics of the saint were shown. He saw they were not Rosalia's. 'They are the bones of a goat', he cried out, not of a woman; and the sanctuary doors were abruptly closed.
    >
    > Frank used to tell of their visit long afterwards to a foreign cathedral where was exhibited a martyr's blood – dark spots on the pavement ever fresh and ineradicable. The professor dropped on the pavement and touched the stain with his tongue. 'I can tell you what it is; it is bat's urine!'

35. Buckland, W. (1820). *Vindiciae Geologicae; or the connection of geology with religion explained*. Oxford.

36. Buckland, *op. cit.* (Note 35), p. 13.

37. *cf.* Hutton's late discovery of granite veins and angular unconformities after his theory was formulated.
38. Curious to learn more about the habits of hyenas, to aid him in his researches. Buckland had one imported from Africa, and it quickly adapted itself to the family menagerie. One evening a supper guest was somewhat disturbed to hear a crunching noise coming from beneath the sofa. 'Do not concern yourself with that', cried Buckland, 'it will only be the hyena eating one of the guinea pigs.'
39. Buckland, W. (1823). *Reliquiae Diluvianae.* Murray, London.
40. Sedgwick, A. (1825). *Ann Philos*, n.s. **10**, 34.
41. Fleming, J. (1826). *Edinb. Philos. J.* **14**, 208.
42. Scrope, G. P. (1825). *Considerations on volcanoes.* Phillips, London.
43. Scrope, G. P. (1827). *The geology and extinct volcanoes of Central France, including the volcanic formations of Auvergne, the Velay and the Vivarais.* 2 vols. Murray, London.
44. Scrope (1827) *op. cit.*, p. 236.
45. The theory might be long discredited, but the name survives in popular parlance, with reference for example to dinosaurs or indeed anything unspeakably ancient.
46. Lyell's modern biographer, L. G. Wilson, gives a full account of his early researches leading to the *Principles of geology*, in *Charles Lyell: the years to 1841, the revolution in geology.* Yale University Press (1972). See also Bailey, E. B. (1962). *Charles Lyell.* Nelson, London.
47. Lyell, C. (1826). *Quart. Rev.* **34**, 518.
48. Rudwick, M. J. S. (1969). Lyell on Etna, and the antiquity of the earth. In *Towards a history of geology* (ed. C. C. Schneer), p. 288. MIT Press.
49. Letter of Lyell to Murchison, 15 Jan. 1829. In Lyell, K. M. (1881), *Life, letters and journals of Sir Charles Lyell*, Vol. 1, pp. 234–5. Murray, London.
50. Rudwick, M. J. S. (1969). Introduction to Lyell's *Principles of geology*, Vol. 1, In *The Sources of Science*, no. 84. Johnson Reprint Corporation, New York and London.
51. Lyell, *op. cit.* (Note 50), p. 165.
52. Lyell, *op. cit.* (Note 50), p. 64.
53. Hutton, *op. cit.* (Note 4), p. 302.
54. Both Hutton's and Lyell's conception was epistemological rather than ontological; the evidence forbids enquiry into beginnings and ends and hence any consideration of the eternal.
55. Laudan, R. (1982). *Studies in History and Philosophy of Science* **13**, 215.
56. The term was introduced by Lyell in Volume 3; the concept is Hutton's.
57. Studer, B. (1827). *Ann. Sci. Nat.* **11**, 5–47.
58. Rudwick (*Palaeontology* **21**, p. 225 (1978) has argued that Lyell's work on Tertiary biostratigraphy was motivated by the hope of establishing a statistical palaeontology which would eventually provide a general chronometer for the whole of the fossil record. Lyell's dream faded quickly, however, as it attracted substantial criticism from his contemporaries. The major faunal discontinuity at the top of the Cretaceous

Chalk probably discouraged him from attempting his pursuit further back in time.

59. Lyell, K. M. *op. cit.* (Note 50).
60. Lyell, K. M. *op. cit.* (Note 49), pp. 271, 273.
61. Whewell, W. (1932), *Quart. Rev.* **47**, 126. Whewell was a formidable figure, a scientific polymath who ended his career as Master of Trinity College, Cambridge. Besides being an authority on earth tides he has fair claim to being the first major historian and philosopher of science, Bacon perhaps excepted. In fact we owe it partly to his influence that the term *science* replaced the hitherto universally used *natural philosophy*.
62. This term was proposed by Rudwick, M. J. S. (1971). In *Perspectives in the history of science and technology* (ed. H. D. R. Duane) p. 209. Norman, Oklahoma.
63. Conybeare, W. D. (1830–1), *Philos. Mag.* **8**, 215, 359, 401; **9**, 19, 111, 188, 258.
64. Sedgwick, A. (1831), *Proc. Geol. Soc.* **I**, 313.
65. Sedgwick, *op. cit.* (Note 64), p. 305.
66. Conybeare, W. D. (1831–2), *British Assoc. Rep.*, 1 and 11, Report on Geology, p. 406.
67. Whewell, W. (1837), *History of the inductive sciences, from the earliest to the present time*, Vol. 3, p. 616. Parker, London.
68. Whewell, *op. cit.* (Note 67), Vol. 2, p. 120.
69. Whewell, W. (1831). *British Critic* **9**, 194.
70. See, for instance, Green, V. H. H. (1974). *A history of Oxford University.* Batsford, London.
71. de la Beche, H. (1831). *A geological manual* (1st edn), p. 32. Truettel and Wurtz, London.
72. This was published as Darwin, C. (1840). *Trans. Geol. Soc.* (ser. 3) **5**, 601. For a reprint of Darwin's collected geological writings see Darwin, C. (1910). *Coral reefs; volcanic islands; South American geology.* Ward Lock, London.
73. Lyell, C. (1850). *Quart, J. Geol. Soc.* **6**, xxxii.
74. Lyell, *op. cit.* (Note 73), p. xlv.
75. Lyell, *op. cit.* (Note73), p. lv.
76. Lyell, C. (1851), *Quart. J. Geol. Soc.* **7**, xxxii.
77. Lyell, *op. cit.* (Note 76), p lxxiv.

3
The emergence of stratigraphy

Generations of British and American geologists have been indoctrinated as students with the view that the English canal surveyor William Smith (1769–1839) was the father of stratigraphy. This has never been accepted, with good reason, on the European continent, where stratigraphy has been perceived as emerging gradually from a long tradition dating well back into the eighteenth century, with the seminal influence deriving from Werner.[1]

It was in the eighteenth century that the systematic study of minerals began in earnest, with four major classes being generally accepted—earths, metals, salts, and bituminous substances. These could be distinguished by their reactions to heat and water, though the means by which they were consolidated posed a major problem. The distinction of 'earths' was especially significant. The action of aqueous fluids, rather than heat, was widely thought to be the key to understanding their formation and transformation, with even siliceous earths being soluble in some circumstances. The beginning of an historical approach to the earth was to be found in the speculative theories of the cosmogenists of the late seventeenth and early eighteenth centuries, against which there was such a strong empiricist reaction several decades later (see Chapter 1). Cosmogenists took scriptural writings as a historical record, though a simple literalism was already out of favour, and an original fluid planet was generally accepted, with the more or less spherical shape being the consequence of rotation. Rocks of the earth's crust were formed by precipitation or deposition from water. By way of explanation, analogies was drawn with chemical experiments. The older 'Primary' rocks were composed largely of siliceous, less soluble earths and the younger 'Secondary' rocks of more soluble earths.

Werner's research represents the culmination of this chemical tradition of mineralogy and cosmogeny. His originality lay principally in his making the time of formation of the rocks, rather than their mineralogy, their most significant character. Although his use of rock formations as unique historical entities, rather than natural kinds, was anticipated by earlier researchers such as Lehman and Füchsel, Werner was the first to propose a system according them a universal significance. Werner's formations (*Gebirge*) have already received some discussion in Chapter 1. They were identified primarily by mineralogical character and secondarily by topographic relief, dip, texture and fossil content, with account also being taken of the principles of superposition first enunciated in the late seventeenth century by Steno. We have seen that Werner's ambitious stratigraphic scheme failed because of the later recognition of igneous intrusion, metamorphism and stratal displacement by tectonic disturbance, and because it became widely recognized by the early nineteenth century that rocks of the same mineralogy could occur repeatedly in·succession. It is important to appreciate, however, that these criticisms came from geologists who staunchly considered themselves to be members of the Wernerian school, and Werner is said to have encouraged his students to study fossils, as perhaps providing a better means of correlation. Whereas Werner himself never used fossils for this purpose, he deserves credit for introducing the concept of correlation, and establishing a global stratigraphy, without which the systematic study of earth history would be quite impossible.

It was indeed fossils, long recognized as the remains or traces of past life, that were to provide the key to correlation, and use of these as biostratigraphic indices led to the launching of the most dynamic geological enterprise of the first half of the nineteenth century. The critical breakthrough was probably achieved by the end of the first decade. The French contribution of Cuvier and Brongniart, based on their study of the strata of the Paris Basin, is well known.[2] Less familiar is the work of the German, Johann Friedrich Blumenbach (1752–1830), who taught at Göttingen, where his most illustrious pupils included Humboldt and Schlotheim. Both Blumenbach and Schlotheim were strong advocates, with Cuvier, of successive extinctions in the fossil record, with the loss from the earth of a series of ammonites, belemnites and land mammals, and of the use of fossils for correlation. Schlotheim made a detailed study of fossil plants in Thuringia in strata now attributed to the Permian, and became convinced that they were all extinct. Blumenbach's three-fold division of the fossil succession, punctuated by organic 'revolutions', was taken a stage further by Cuvier, who advocated more frequent and violent episodes of organic turnover, introducing catastrophist overtones that had never been present in Werner's work. In England Thomas Webster quickly applied Cuvier and Brongniart's Paris Basin scheme to the Cretaceous and Tertiary of the Isle of Wight, thereby proving that fossils were at least in this instance more reliable for correlation than mineral character.

Where does William Smith stand in relation to these developments? Smith was for a long time an isolated figure within the geological community in Britain, let alone the continent, and his belatedly-acknowledged most significant contribution was perceived as being in geological map making. The extent to which he relied predominantly on fossils for correlation in mapping geological formations has been called into question by Laudan.[3] Thus he apparently confused the Magnesian (=Permian) and Mountain (= Carboniferous) limestones and put the rocks of the Blackdown Hills (= mid Cretaceous Upper Greensand), Weald (= Lower Cretaceous Wealden) and North York Moors (= mid Jurassic) into the same formation, in his pioneer geological map of England and Wales. The formations have similar rock types, or lithofacies, but different fossils, or biofacies. Not until quite late in his life did Smith become a strong advocate of the use of fossils for biostratigraphy, probably under the influence of his palaeontology-promoting nephew, John Philips. In 1831 Smith was dubbed by Murchison 'the father of English geology', but most British geologists learned their mapping from the continental tradition, well exemplified by the early maps of Buch, Elie de Beaumont, and Omalius d'Halloy, and there is no evidence that he had much influence in the critical first two decades of the nineteenth century. In all likelihood he discovered the stratigraphic value of fossils quite independently of Cuvier and Blumenbach; unlike them he did not develop any theory of extinction. Indeed, it is hard to think of another geological pioneer who worked so independently of any kind of theory.

The end of the Napoleonic wars allowed more foreign travel and by the 1820s a number of tables of international stratigraphic correlation had been published. Humboldt was a major driving force behind this endeavour, while Buckland was the leading English figure.[4] His stratigraphic column for the British Isles and Europe, published in 1821, showed eleven divisions: (1) Transition Limestone and Grauwacke (equivalent to Lower Palaeozoic), (2) Old Red Sandstone, (3) Carboniferous Limestone, (4) Coal Measures, (5) New Red Conglomerate, (6) New Red Sandstone and Magnesian Limestone, (7) Oolite or Jura Limestone, (8) Chalk and Greensand, (9) Tertiary, (10) Diluvium, (11) Alluvium. Buckland was also a palaeoecological pioneer. He argued that cycadeous trunks in a Jurassic soil bed in the Isle of Purbeck, Dorset, signified a former tropical climate, and a black substance found associated with a belemnite in the Lias of Lyme Regis was interpreted as the fossil ink sac of an ancient cuttlefish. He took delight in having fossil specimens drawn in their own reactivated ink, while his recognition and description of ichthyosaur coprolites from the same deposits evidently appealed to his rather coarse sense of humour. His work on footprints in the Scottish New Red Sandstone marks the beginning of ichnological studies.[5]

While from the beginning of the 1820s geologists paid increasingly close attention to fossils, formations remained the primary units of description, and not everyone was convinced of the importance of palaeontology in correlation. The principal sceptic in England was one of the founders of the Geological Society of London, George Greenough, whose geological map of England and Wales – essentially a collaborative effort – was published in 1820; it displayed much more accuracy in most areas than Smith's solo effort of 1815. At the other extreme Lyell had attempted to correlate the Tertiary entirely by means of fossils, using the percentage of extant species as the key criterion (see Chapter 2).

The challenge of the ancient slaty rocks

By the 1830s a substantial number of regional stratigraphic memoirs had been published and the detection and analysis of folds, faults, and unconformities had become routine. De la Beche's popular *Geological Manual*, published in 1831, gives a succession differing little from that of Buckland, published a decade earlier. Correlation with the standard Wernerian scheme is shown below

de la Beche	*Wernerian*
Modern	Alluvion
Erratic blocks	Diluvium
Supracretaceous	Tertiary
Cretaceous Group	
Oolitic Group	
Red Sandstone Group	Secondary
Carboniferous Group including Old Red Sandstone	
Grauwacke Group	Transition
Primitive	Primary

The greatest successes had been achieved with the Secondary strata, which in Great Britain started with the Old Red Sandstone and Carboniferous Limestone. This is because there was a clearly defined stratigraphic sequence in generally coherent rocks with abundant diagnostic fossils. The Primitive or Primary rocks comprised the complex igneous and metamorphic basement and were to remain stratigraphically beyond the pale until this century, and it was generally agreed in the 1830s that the biggest challenge to stratigraphic analysis was posed by the Grauwacke or Transition Rocks, which were in general sparsely fossiliferous and usually tectonically disturbed, and hence difficult to interpret. Accordingly no stratigraphic sequence had yet been elucidated.

The two leading geologists to take up the challenge were Sedgwick and Murchison. Sedgwick has figured prominently in earlier chapters but Roderick Murchison (1792–1871) requires some introduction here. Independently wealthy, with a family estate in Scotland (though he lived in England nearly all his life), he had devoted himself to the life of a country gentleman after an unsuccessful military career. Though much given to fox hunting and socializing he quickly acquired a passion for geology, joining the Geological Society of London and attending lectures at the Royal Institution. One of his principal mentors was Buckland, whose influence showed up in the close attention he paid to fossils. By the mid 1820s he had established that the Brora coal of north east Scotland contained plants similar to those in the 'Oolitic' strata of Yorkshire, which confirmed to him the primary importance of using palaeontological criteria for correlation. In contrast, Sedgwick was much more concerned with the physical character and structural disposition of rocks, and considered the establishment of reliable sections across a particular terrain to be of paramount importance. Much influenced by Elie de Beaumont's work on mountain building, he introduced the term *strike* into English geological usage (from the German *Streichen*), first published the term *synclinal*, and developed criteria for distinguishing bedding from cleavage. Both Murchison's and Sedgwick's preferred types of approach were to prove of great value in elucidating the Transition rocks, but their differences in emphasis and outlook were also to contribute to what eventually developed into one of the most acrimonious geological debates on record.

However, as will be shown, the Cambrian-Silurian controversy[6] was not just between Sedgwick and Murchison, and they enjoyed a friendly collaborative partnership for a considerable time. This collaboration extended to investigation of the rocks of south west England, giving rise to another major controversy concerning the Devonian.[7] Though the story of these two controversies involves what many may consider excessive parochial detail, the essential issues were anything but parochial, and had a vital influence in what appeared to contemporaries as the most central of geological endeavours through much of the nineteenth century, namely the satisfactory unravelling and correlation of the stratal sequence.

Sedgwick and Murchison's early research in Wales

Sedgwick and Murchison had done fieldwork together in the 1820s, from Scotland to the Alps, and both decided to undertake in 1831 a preliminary investigation into the Grauwacke (or Greywacke) of Wales and the Welsh borders, hitherto virtually *terra incognita* geologically speaking. They agreed to cover different territories, Murchison Shropshire and areas to the south west, Sedgwick North Wales, from the Berwyn Range westwards to Caernarvon. Murchison had the good luck to discover a series of richly fossiliferous passage beds beneath the Old Red Sandstone into what could unequivocally be ascribed to the Grauwacke. Being an astute and ambitious man, he rapidly perceived the rich potential of his selected territory for future research. Sedgwick, who was accompanied for a few days by Charles Darwin, was less fortunate, because there was a major unconformity in North Wales below the Carboniferous Limestone, and the underlying rocks were only sparsely fossiliferous. He was in consequence forced to utilize a marker horizon well down in the Grauwacke Group and was unable to make a link with the Secondary rocks.

Both geologists returned to their chosen field areas for more substantial work in the following summer. Murchison's classification of the Grauwacke or Transition strata varied a little in the first few years but eventually stabilized to the following formations:

Ludlow Rocks
Wenlock Limestone
Caradoc Sandstone
Llandeilo Flags

Sedgwick found the key to his interpretation in what he termed the Bala Limestone, a 6–10 metre thick unit on the western slopes of the Berwyns, which was one of the few distinctive fossiliferous horizons. Overlying the Bala Limestone was a series of poorly fossiliferous slates followed by 'upper calcareous greywacke', and underlying it to the west a calcerous series in turn underlain by the slates and associated igneous rocks of Snowdonia, similarly very poorly fossiliferous. Primary basement was recognized in Caernarvon and Anglesey. Clearly Sedgwick had been working in what could be ascribed to Lower Grauwacke and Murchison to Upper Grauwacke and the question naturally arose as to how Murchison's lower strata could be related to Sedgwick's upper strata. Initially they thought they were studying strata at different horizons and in 1834 undertook a joint tour with the object of finding a collaborative boundary. The tour began with an examination of Murchison's best sections, and then moved on to the Berwyns, which in retrospect became the key to the Cambrian-Silurian controversy.

The Meifod Limestone of this region was ascribed by Murchison to the Caradoc because of the general similarity of its fossils. To the north west of its outcrop were slates containing the trilobite *Asaphus buchii*, which was a key marker for the Llandeilo. The principal question to resolve concerned the stratigraphic position of the Bala Limestone. Though Murchison was struck by the similarity of Bala, Meifod and Caradoc fossils neither he nor Sedgwick apparently considered treating the Bala and Caradoc as stratigraphic equivalents, and a mutual decision was made to place the Bala Limestone below the Llandeilo.

The results of the Berwyn traverse were subsequently expanded by tracing the agreed mutual boundary throughout the rest of Wales, Sedgwick concentrating on the north, Murchison on the south, utilizing a variety of criteria, such as changes in lithology and strike, the amount of slaty cleavage, and fossil content. At the 1835 meeting of the British Association for the Advancement of Science the two collaborators announced the creation of two new geological systems, the Silurian and Cambrian. Murchison's Silurian System, named after an ancient British tribe of the Welsh borders that had harassed the Romans, embraced all formations between the Old Red Sandstone and the Grauwacke proper, which was ascribed to the Cambrian, a term derived from the Roman name for Wales.

SILURIAN	Upper	Ludlow Rocks
		Wenlock Limestone
	Lower	Caradoc Sandstones
		Llandeilo Flags
CAMBRIAN	Upper	Higher beds of Berwyns
		South Wales States
		Bala Limestone
	Middle	Higher mountains of Caernarvon and Merioneth
		Snowdon Beds
		Anglesey and SW Caernarvon schists and gneisses
	Lower	(no fossils)

Sedgwick subsequently (1838) excluded the 'Lower Cambrian' metamorphics and put them in a 'Primary Class' of unstratified rocks.

Before we pursue further the Cambrian-Silurian story for Wales it is desirable to digress to south west England. Both Sedgwick and Murchison had good reason to be interested in the geology of this region. Sedgwick had noted the overall similarity of the Upper Cambrian and Lower Silurian rocks of Wales and had, as a follower of Elie de Beaumont, sought to use differences in strike as a means of distinguishing the two systems. His Welsh Cambrian being on the whole poorly fossiliferous, he entertained high hopes of finding a really distinctive Cambrian fauna in the limestones of Devon. Murchison had become firmly convinced by his Silurian work that it was futile to look for coal in the Grauwacke, and in consequence was highly provoked when just such a claim was made for Devon.

The Devonian controversy

Henry de la Beche had for some time been interested in producing a geological survey of south west England, much of which was portrayed simply as Grauwacke on Greenough's map. He derived a considerable income from his family estate in Jamaica until in 1831 he suffered a drastic decline in his fortunes. Thus he was subsequently forced to look elsewhere for financial support, and attempted to get his geological research subsidized by the government. This proved difficult compared with the generous support provided for the official geological survey of France. Fortunately the wealthy Greenough acted as a patron and protector.

At the December 1834 meeting of the Geological Society, de la Beche had a letter read (he was too impoverished to travel to London himself) in which he reported the discovery of plants of Coal Measures type in strata in North Devon attributed to Grauwacke. This claim stimulated Murchison into making a vigorous attack on de la Beche, whom he castigated for placing reliance on rock types reminiscent of the old discredited Wernerian geognosy. The term Grauwacke should not be used both as a rock type and as a stratigraphic unit. The plants signified Coal Measures, which implied that de la Beche must have overlooked a major unconformity. Whereas Lyell supported this view, Greenough rose to de la Beche's defence, observing acidly that Murchison had never visited the area in question whereas de la Beche knew it well. On learning of the attack de la Beche was infuriated, and he sought Sedgwick's support against Murchison and Lyell, whom he evidently considered mere armchair geologists. Sedgwick responded sympathetically, playing down the role of fossils *vis-à-vis* the evidence of superposition given by actual sections. In correspondence Sedgwick drew Murchison's attention to the fact that his argument was based on purely negative evidence; no-one knew what the plant world looked like in Grauwacke times. John Phillips, entering the fray, thought that it was quite possible that a flora of so-called Coal Measures species could have flourished locally in particular ecological conditions during the Grauwacke period, long before it was able to spread widely to give rise to the characteristic Coal Measures flora.

No doubt much of de la Beche's anger was allied to anxiety because of the threat to his reputation while he was still a supplicant for a government-supported post. Fortunately for him his livelihood was secured in the middle of 1835 by his appointment as head of the new geological branch of the Ordnance Survey, a position that was to lead very quickly to his becoming the first director of the Geological Survey. Only a few months after this Sedgwick shifted his allegiance in the argument from de la Beche to Murchison, having been impressed by the fact that de la Beche had coloured some Culm-type plant-bearing rocks, similar to those in Devon, as Grauwacke in his geological map of Pembrokeshire in South Wales. To the north of and stratigraphically below these rocks there existed undoubted Grauwacke, which was deceptively conformable but had a different strike; herein perhaps lay the error.

It was by now imperative that both Sedgwick and Murchison should visit Devon themselves, which they did in the summer of 1836. They accepted the older strata of North Devon as Lower Silurian and therefore expected to find evidence of a major unconformity below the plant-bearing Culm, which was thought to be true Coal Measures. However, no such evidence was found. They observed black limestones with the bivalve *Posidonia* in the long part of the Culm series, and thin coals higher in the succession. Contrary to what de la Beche had claimed in his report, there was no slaty cleavage in any of the Culm rocks. Evidently the Culm was uppermost in the sequence, not within the Grauwacke. Subsequently Sedgwick and Murchison transferred their attention to South Devon, where the fossiliferous limestone around Plymouth and elsewhere was thought to be located well down into the Cambrian. A rapid north–south traverse across central Devon revealed a broad trough containing Culm, stratigraphically equivalent at least in part to the Coal Measures, flanked to the north by Cambrian plus a little Lower Silurian and to the south by thick Cambrian only. They thought that a sub-Culm unconformity must be present though not observed.

At the British Association meeting at Bristol in 1836 Sedgwick and Murchison reported their findings, to the evident surprise of de la Beche, who objected to having a paper sprung on him without warning, something he considered ungentlemanly. He accepted that the Culm must be uppermost in the sequence but refused to concede it as Carboniferous. In defence of his position, that all the Devon rocks were pre-Carboniferous, perhaps even pre-Silurian, he pressed home the point that no unconformity had been discovered. Phillips, however, supported Sedgwick and Murchison's view that the Culm was really Carboniferous, because *Posidonia* was a fossil characteristic of the Mountain (or Carboniferous) Limestone of Yorkshire. The clastic

Culm could indeed be older than the Coal Measures, as implied by the fossil, because he had shown that the Mountain Limestone passes laterally into coal-bearing clastic deposits in Scotland.

Following the British Association meeting Lyell, in his Presidential Address to the Geological Society, noted that as far as he was concerned the clinching point of the argument was that de la Beche had failed to demonstrate that the plant-bearing Culm strata were anywhere intercalated with strata containing undisputed Transition fossils. The controversy was however by no means settled. Thus there was a problem about the occurrence of fossil plants in strata that Murchison had assigned to the Lower Silurian. According to him there were no authentic records of land plants as old as the Silurian, and de la Beche used this evidence to support an upward passage of the plant-bearing beds without a break into the Culm. Sedgwick and Murchison's paper was finally read, after much procrastination by Sedgwick, at the Geological Society meeting in June 1837. The Upper Cambrian was well developed in Devon, with Lower Silurian being present in the uppermost North Devon strata, an attribution based on rock types rather than fossils. The Lower Silurian attribution was not a confident one and the problem of the plants was glossed over. The Culm, from which goniatites were now recorded, was considered to be equivalent to the Mountain Limestone and Coal Measures.

As the controversy entered its fourth year the spotlight turned to the limestones in South Devon and an important new figure emerged, namely Robert Austen, a very capable local geologist, who had assisted Sedgwick in the field in 1836. In a paper read to the Geological Society, Austen claimed that in South Devon the Culm unconformably overlay 'Transition' rocks, in which the Devon limestones were placed. In insisting that the stratal sequence was continuous and conformable in North Devon, Austen agreed with de la Beche, but in attributing the Culm to the Carboniferous he supported Sedgwick and Murchison. The most important new idea he put forward was that his 'great limestone' of south east Devon was equivalent to the Mountain Limestone. In coming to this conclusion he was influenced by the work of the palaeontologists William Lonsdale and James de Carl Sowerby, who were impressed by the approximate similarity of the Devon fossils to those in rocks of similar facies in the Carboniferous. On the other hand, Silurian faunas were distinctly different from the Carboniferous. Austen's conclusion about the age of the Devon limestones was very upsetting for Sedgwick, because it took away his 'magnificent development' of the Upper Cambrian, but it did justice both to the unbroken limestone sequence in South Devon and expert opinion concerning the fossils he had found therein. Sedgwick had in fact failed to characterize the Cambrian biologically as Murchison had managed to achieve for the Silurian.

Whereas de la Beche had assigned the whole of the Culm to the Grauwacke, Sedgwick differed from Murchison in being prepared to concede that at least the lower part did belong there, and began to backtrack on the Culm–'Transition' unconformity, though he still considered most of the older rocks to be Cambrian. He was evidently perplexed by the limestone faunas. One of his Cornish localities, believed to be amongst the most convincing, had yielded the cephalopod *Clymenia*, chosen by Lyell to exemplify the Cambrian in his *Elements of geology*. The Cornish fauna included, however, other fossils indistinguishable from the faunas of the limestones in South Devon.

So far no attempt had been made to correlate the Devon fossils with those found elsewhere, but this commenced when Edward de Verneuil, a Parisian lawyer and fossil collector, showed Austen his collections from the Eifel district of the Rhineland. Austen was impressed by their close resemblance to those in the Devon limestones but Verneuil insisted that the Eifel limestone must be Silurian. Interest now arose in the successions of Belgium and the Boulonnais, in the north western extremity of France.

The start of 1839 marked the publication of two important works.[8] In de la Beche's long-awaited comprehensive report on the geology of south west England he correlated the Culm with the Upper Grauwacke, which implies an approximate equivalence to Murchison's Silurian, but with both being considered merely local deposits. He persisted in his view that *all* the older rocks belonged to the Grauwacke, but was prepared to concede that the sequence might extend up to the equivalent of the Old Red Sandstone. Of much greater international significance was Murchison's monograph on the Silurian system, dedicated to Sedgwick, in which he stressed that fossils are the key to correlation. The Old Red Sandstone was elevated to a system in its own right rather than being a mere base to the Carboniferous. It would, he claimed, be a worthwhile strategy to seek a transitional fauna that would blur the sharp contrast between the Silurian and Carboniferous.

Murchison was clearly attracted to the notion that all the old rocks of south west England might be stratigraphically equivalent to the Old Red Sandstone, particularly because it would relieve the embarrassment of being unable to find a sub-Culm unconformity. Because, however, there were no fossils in common, such a correlation could only be achieved at the cost of abandoning the full rigour of his insistence on the critical importance of fossil evidence. Though

Sedgwick had reservations, feeling that his Cambrian was threatened, he was persuaded by Murchison to submit a joint paper for rapid publication[9] in which the Devonian System was first proposed, with the striking assertion being made that 'the oldest slaty and arenaceous rocks of Devon and Cornwall are the equivalents of the old red sandstone'. Fossil molluscs, corals, brachiopods, and trilobites of the South Devon limestones were of mixed Silurian-Carboniferous character. Among other advantages of this proposal, Murchison's claim that the Silurian predated any land plants could be salvaged.

Reaction to this proposal was somewhat varied. While Buckland was sympathetic, Greenough expressed hostility to what he saw as the grandiose creation of a new stratigraphic system on purely local grounds. De la Beche raised no strong objections, however, while Austen evidently resented the lack of acknowledgement of his important work by what he thought of as a Geological Society clique.

Murchison was acutely conscious of the fact that clear evidence was lacking of Silurian strata underlying Devonian. It was obviously necessary to undertake investigations in strata of an equivalent age on the continent. Thus in the summer of 1839 he set off with Sedgwick for fieldwork with local geologists in Germany and Belgium. The results were initially somewhat confusing. Apparent Wenlock fossils were found in the Rhineland and the Ardennes but there was no sign of a Ludlow–Devonian transition. Sedgwick stayed on in Belgium after Murchison's return to England and was persuaded that the Ardennes rocks must be of Cambrian age. Moving on to the Rhineland, Sedgwick came to the conclusion that what had originally been taken as Devonian on both sides of the Rhine must in fact be Lower Silurian. Murchison returned to the continent for the meeting of the Societé Géologique de France at Boulogne. He accepted Verneuil's correlation of the Silurian for the Boulonnais, which left virtually no room for the Devonian.

Verneuil's correlations had serious implications for the Rhineland because the Devonian there had almost vanished. Murchison had turned most of the 'Grauwacke' there into Devonian, only to find himself obliged by the fossil evidence to convert it back into Silurian. Not surprisingly, Sedgwick completely lost confidence in the Devonian interpretation and considered all the pre-Culm rocks in south west England to be either Lower Silurian or Cambrian. However, at the turn of the year 1839–40 Murchison's confidence in the new system was restored by the evidence of fossils, because those found beneath undoubted Carboniferous were decidedly not Silurian forms, and both Lonsdale and Sowerby identified supposed Silurian fossils in the Ardennes as Devonian. Lonsdale had found the key clue in Austen's fossil collection from South Devon. He had agreed with Sowerby that the brachiopods and molluscs resembled Carboniferous forms but thought that the corals were more like Silurian ones, and this mixed character first led him to suspect that the rocks which contained them were probably equivalent to the Old Red Sandstone. This was supported by the identification by Louis Agassiz, the leading expert on the subject, of the distinctive Old Red Sandstone fish *Holoptychius* in Belgian formations rich in 'Devonian' shells and corals.

Leading European geologists such as Elie de Beaumont, Humboldt, and Buch supported Murchison and decisive confirmation came shortly afterwards by the discovery, in Russia, near Lake Ladoga, and in America in New York State, of characteristic Old Red Sandstone fish in association with marine Devonian shells. By the summer of 1841 only a few marginal figures in the controversy rejected the Devonian interpretation, and by the middle of the decade research by German geologists in the Rhineland had led to the virtual elimination there of the Silurian. For a time 'Rheinisch' competed with 'Devonisch' among the Germans but the term Devonian was soon adopted because of historical precedence and has been in general use ever since. Nevertheless it is in the Rhineland and Belgium that a full sequence of Devonian faunas was worked out thoroughly and subdivisions defined, with correlations being made from the Ural Mountains to the eastern United States.

Development of the Cambrian-Silurian controversy

The second volume of Murchison's *Silurian System*, devoted to palaeontology, demonstrated a clear difference in fossil content between the Upper and Lower Silurian, though an overall unity was considered to be present. Characterization of the Cambrian was explicitly left to Sedgwick, with the two friends continuing to respect their agreed territorial limits. Sedgwick was confident that a vast sequence of fossiliferous rocks extended down from the Silurian, and both he and Murchison, as geological progressionists, considered it extremely unlikely that a single fauna could have inhabited the earth for such a long time. In 1841 Sedgwick expressed the hope of achieving a 'progressionist' method of correlation, marked by the gradual introduction of Silurian fossils through time. This idea was treated coolly by the Geological Society fellows who, in tune with their expressed bias towards empiricism, pointed out the need for more field evidence. Thus Sedgwick set out to North Wales in 1842 in the company of a young palaeontologist, John Salter; they failed to find any significant difference between the fossils of the lowest and highest beds.

Meanwhile Murchison's monograph on the Silurian had been well received internationally, and he was encouraged to extend his system geographically and stratigraphically because of the failure to find a distinctive Cambrian fauna. Buch indicated to him that there occurred in the northern part of the Russian Platform a series of tectonically-undisturbed richly-fossiliferous 'Transition' strata, and in 1840 Murchison travelled extensively in Russia with Verneuil and von Keyserling. Besides obtaining confirmatory evidence for establishment of the Devonian, he laid the foundations for a new system, the Permian. By 1842, when he became President of the Geological Society, his personal reputation was assured, leading a few years later to the award of the Copley Medal of the Royal Society and a knighthood. His recently purchased home in one of the most fashionable parts of London became a magnet for that City's social and scientific elite. None of this would have done much to diminish his already considerable sense of self-importance. His two presidential addresses have indeed both a triumphalist and imperialist tone, because he thought that global biostratigraphy would be modelled on British strata. He noted that the overwhelming mass of evidence accumulated since publication of *The Silurian system* demanded the complete abandonment of the Cambrian as an internationally valid term. Through the rest of the decade Sedgwick was one of the few dissenters to this view, stressing that the 'break' between the Upper and Lower Silurian would prove more important than that between the Upper Silurian and Devonian.

Other challenges to Sedgwick came in the early 1840s from the fledgling Geological Survey of Great Britain and an independent investigator, Daniel Sharpe. Andrew Ramsay, one of the brightest of the Geological Survey geologists, made the significant discovery that Murchison's Lower Silurian slates extended north westwards to cover most of South Wales. He and his colleagues found characteristic Caradoc and Llandeilo fossils deep within the supposed Cambrian, and Sedgwick and Murchison's original Cambrian-Silurian boundary had to be abandoned. Sharpe had the temerity to contest Sedgwick's structural interpretation in North Wales, arguing that the Berwyn strata lay below the Bala Limestone, not above. The main implication of this controversial view was that the Berwyns formed the type area of a 'Cambrian' completely devoid of fossils.

Sedgwick responded with vigour to the attack by Sharpe, who was widely perceived to be a rather brash upstart. The dispute hinged crucially on the dip direction of the Bala Limestone, which did not prove an easy matter to resolve, but it is clear that Sedgwick was forced to defend his position strongly, because Sharpe's interpretation would imply that all the North Wales rocks should be put into the Silurian. To his dismay, Murchison was disinclined to give him the support that he felt was due, which suggested implicitly that he put the blame on him for any errors the two had made in fixing the position of the Cambrian-Silurian boundary. An active debate between Sedgwick and Murchison continued in private correspondence, but contemporary published papers give little idea of the heat of the debate. In 1843 Sedgwick had abandoned the term Cambrian in favour of Protozoic but three years later revived it, perhaps as a consequence of the provocation he was still receiving from various quarters. However, it still lacked a fauna of its own, being merely characterized by the progressive appearance of Silurian fossils. The resurrection of the Cambrian, even in limited form, posed a threat to the integrity of Murchison's Silurian, and Murchison vigorously disputed the most critical of Sedgwick's claims, such as placing the South Wales slates in the Upper Silurian, creating a series of passage beds linking the Wenlock with the Caradoc and Llandeilo, and constructing a Cambrian 'group' containing mainly Lower Silurian fossils but lacking *Asaphus buchii*.

By the beginning of 1847 Sedgwick and Murchison had established their basic positions that were to change little subsequently. Murchison so defined his Silurian that it would include 'Cambrian' fossils subsequently discovered, an attitude that received general support. Such support was actively solicited by intensive lobbying because he was alarmed that adoption of Sedgwick's Cambrian would severely attenuate his Silurian. Through the course of the 1840s Sedgwick had become increasingly isolated, and had shifted his ground on a number of occasions; for example, on the extent to which Cambrian and Silurian faunas differed, with passage beds being proposed to replace a widespread unconformity. He believed that by using fossils in the passage beds to incorporate all the underlying Cambrian rocks into the Silurian, Murchison had violated the basic principles of geological nomenclature.

The course of the controversy was significantly affected by publication of the Geological Survey 1 inch/mile maps of North Wales. This meant effectively that Sedgwick's and Sharpe's work was superseded, though Sedgwick's structural interpretation was substantially vindicated. On the other hand the Survey followed Murchison's classification.

In an attempt to resolve the controversy a meeting of the Geological Society was held in 1852, with a lively debate lasting until midnight.

Sedgwick read a paper in which he finally accepted the Survey succession in Wales, in particular uniting the Bala and other limestones to the same horizon, though he continued to repudiate their nomenclature. He regretfully acknowledged that he and other gentlemen geologists were no longer able to compete with the professional expertise of the Survey teams, and abandoned for good his progressionist scheme. In the second part of his paper he launched into a vehement assault on Murchison, who he accused of a whole series of earlier mistakes. 'His nomenclature was premature and his base-line was sectionally wrong; and, so far from leading to discovery, it retarded the progress of Palaeozoic geology for, I believe, not less than ten or twelve years.'

Always an emotional and moralistic man, Sedgwick was so convinced that right was on his side that he distorted the truth and manifestly lost all pretence at objectivity. His bitterness towards his former close companion can perhaps most plausibly be attributed to the disappointment of an old man who had wished the erection of the Cambrian to be seen as his crowning achievement. Sedgwick's angry polemic raised a storm of protest and led to a widespread suspicion that he had gone either senile or mad. From his position of comparative strength, all Murchison had to do was make a calm reply, expressing regret that Sedgwick had not pursued palaeontology further, using Survey maps and sections to support his claims, and noting the widespread international acceptance of his own system. Poor Sedgwick must have been left feeling totally beleaguered, at the nadir of his fortunes.

All was not yet lost, however. Sedgwick's palaeontological assistant at Cambridge, Frederick McCoy, claimed on the basis of fossil discoveries that the Caradoc embraced two different faunas, one with Lower, and the other with Upper Silurian affinities. This claim provoked a field trip by Sedgwick and McCoy in the summer of 1852, to examine the best sections in Wales and the Welsh borders; they had their greatest success at May Hill. Following this, in November Sedgwick read a paper at the Geological Society, in which the Carodoc was divided into a lower Caradoc Formation, attributed to the Cambrian, and an upper May Hill Sandstone, placed in the Silurian. Similar discoveries were claimed in Central Europe and America, suggesting a more important faunal break than between the Silurian and Devonian. While the audience seemed rather sceptical, the Survey was much disturbed because a revision of their mapping seemed necessary, if Sedgwick's proposal was correct. A good example is the classic Onny River section in Shropshire, where one could evidently demonstrate a slight angular unconformity between the 'Pentamerus Beds' (equivalent to the May Hill Sandstone) and the Llandeilo and Bala Beds (equivalent to the Caradoc). Murchison and the Survey had missed this because none had sought it, their minds being unprepared.

Many similar examples were demonstrated within the next few years, much to the embarrassment of the Survey; there was evidently a widespread unconformity within the 'Caradoc'. The American sequences studied by Henry Darwin Rogers revealed a similar abrupt break. Murchison, by now (1855) appointed Director of the Geological Survey, called the May Hill Sandstone Upper Caradoc, incorrectly implying affinities with the Lower Silurian, and irritated even his subordinates by his refusal to concede ground. In a new edition of *The Silurian system* he proposed the name Llandovery for the Pentamerus Beds and May Hill Sandstone.

By 1856 nearly all British geologists agreed that the Upper-Lower Silurian break was one of the biggest in the stratigraphic record, which was seen as a victory for Sedgwick. While this did not lead to acceptance of the version of the Cambrian favoured by Sedgwick and his supporters it did at least demonstrate how unwise it would be to classify all the older fossiliferous rocks as Silurian. Following Sedgwick's November 1852 paper on the splitting of the Caradoc, the Geological Society Council refused to print any more on either side of the dispute, so for the remainder of the decade Sedgwick and his supporters took their case to the British Association, whose meetings became a forum for Sedgwick's anti-Murchison crusade. The distinguished naturalist Edward Forbes spoke for many when he expressed concern that the bad publicity arising from the controversy would lead the uninformed to wrong conclusions, allowing those intellectually disreputable figures, the Scriptural geologists, to cast doubt on the whole stratigraphic enterprise; 'The controversies of naturalists are seldom creditable to the popular reputation of science'. One cannot resist pointing out parenthetically an obvious comparison with the attempt by creationists in North America to exploit disputes between evolutionists within the last couple of decades.

The British Association in its own turn began to apply censorship, such that Sedgwick began to feel like a martyr. A man of great integrity and passionate intensity, compared by Forbes to Don Quixote, with McCoy as Sancho Panza, his sense of righteous indignation led him into self-imposed exile from the metropolitan geological community, and withdrawal from any pretence of friendship with the authoritarian Murchison, whose considerable powers of patronage he could never match.

Significant progress in the dispute arose from a massive research

Table 1   *Alternative classifications for the British Lower Paleozoic in the latter part of the 19th century. After Secord 1986, Fig. 9.4.*

Author	1	2	3	4	5	6	7	8	9
SEDGWICK 1855	Silurian		Upper Cambrian		Middle Cambrian				Lower Cambrian
MURCHISON 1859	Upper Silurian		Lower Silurian (Primordial Silurian)						Cambrian
GEOLOGICAL SURVEY 1866	Upper Silurian		Lower Silurian						Cambrian
JUKES 1857	Upper Silurian		Cambro-Silurian						Cambrian
PHILLIPS 1855	Upper Silurian		Lower Silurian				Cambrian		
LYELL 1865	Upper Silurian	Middle Silurian	Lower Silurian				Cambrian		
LYELL 1871	Upper Silurian		Lower Silurian				Cambrian		
HICKS 1874	Upper Silurian	Middle Silurian	Lower Silurian				Cambrian		
LAPWORTH 1879	Silurian			Ordovician				Cambrian	
Principal formations	1	2	3	4	5	6	7	8	9
	Ludlow	Wenlock	Upper Llandovery = May Hill / Lower Llandovery	Bala = Caradoc Sandstone	Llandeilo	Arenig	Tremadoc	Lingula Flags	Longmynd

effort of ancient Bohemian strata and their fossils by the expatriate Frenchman Joachim Barrande, which led to a gigantic publication over several decades.[11] Barrande distinguished a succession of three faunas. The youngest was held to be equivalent to the Upper Silurian, the intermediate to the Lower Silurian, and the oldest, termed Primordial, which contained the distinctive trilobite *Paradoxides*. This fossil was subsequently found in Pembrokeshire, South Wales, and a number of other 'primordial' fossils were found both here and in North Wales. Thus Barrande tentatively correlated the Lingula Flags with the Primordial, the Tremadoc Slates being added later. Further such fossils soon turned up elsewhere in Europe and the United States (in the Potsdam Sandstone). Of a series of classifications of Cambrian and Silurian rocks proposed in mid-century, that of Lyell proved the most influential, with the Cambrian-Lower Silurian boundary being placed between the Tremadoc and Arenig, and the Lower-Upper (and Middle) Silurian boundary between the Bala/Caradoc and Llandovery (Table 1).

Following the deaths of Murchison, Sedgwick, Lyell and Phillips in the early 1870s a younger generation of geologists continued to engage in controversy on the most suitable classification. The solution eventually adopted universally owes its origin to Charles Lapworth, whose detailed work on graptolite zonation in the Southern Uplands of Scotland marked a considerable biostratigraphic advance. To avoid the continuing contention between rival Sedgwick and Murchison supporters he replaced 'Lower Silurian' by Ordovician, adopting the name of a Romano-British tribe that had lived in the Bala district of North Wales.

Every geologist will at last be driven to the same conclusion that Nature has distributed our Lower Palaeozoic rocks in three subequal systems, and that history, circumstance, and geologic convenience have so arranged matters that the title here proposed for the central system is the only one possible.[12]

The Geological Survey took some convincing, and only adopted the Ordovician two decades later following the retirement of its director Sir Archibald Geikie. Not until 1960 was the Ordovician officially approved by the Stratigraphic Commission of the International Geological Congress.

The triumph of stratigraphic palaeontology

Intriguing though the story of personal rivalries may be, more general issues were at stake. The history of these two controversies indicates that conversion to the consensual view that fossils were the sole reliable key to extensive correlation was a prolonged and haltering process as far as the older rocks were concerned. Despite a few setbacks Murchison was on the right track from the start, whereas Sedgwick suffered some eclipse of his reputation, partly because he tended to play down the importance of fossils in favour of physical stratigraphy and structure.

Further problems arose because of facies variations, and development of the concept of facies proved far more elusive than correlation. Constant Prévost deserves credit as a pioneer. He recognized that the Wealden of England must be the non-marine equivalent of the marine Neocomian of south eastern France and Switzerland because they were both sandwiched between the Jurasic and the Chalk, and later pointed out that the German Muschelkalk must be represented elsewhere by clays and sands. Probably the greatest credit however should go to the Swiss geologist, Armand Gressly, who worked in great detail on the younger Jurassic rocks of the Swiss Jura and solved the definition of facies still usable today. He demonstrated in his memoir[13] that each of his facies reflected a different environment of deposition, and emphasized (over half a century before Walter proposed his law of correlation of facies) that the vertical and horizontal relations of facies must be compatible in terms of the environments each represents.

As regards contemporary British geologists, John Phillips appears to have been the one with the best appreciation of facies (though he never used the term) based primarily on his research on the Carboniferous Limestone and its lateral equivalents. He was also a pioneer in seeking to achieve an entirely palaeontological classification of the whole fossiliferous sequence of strata, what we now call the Phanerozoic. In proposing the terms Palaeozoic, Mesozoic and Cainozoic[14] he was much influenced by Lyell's treatment of Tertiary faunas and Adolphe Brongniart's palaeobotanical studies in emphasizing the importance of analysing large numbers of species in order to rule out local ecological as opposed to temporal effects. In other words he had progressed far beyond his uncle William Smith's recording of a few key indices, and at times he berated his contemporary, Murchison, for citing too few fossils.

A further important mid-century advance was the erection of stratigraphic stages by the French geologist Alcide d'Orbigny. D'Orbigny perceived that a consequence of Gressly's facies was that stratigraphy needed to be released from the vagaries of formation names based on local facies bodies. As a disciple of Cuvier he believed in repeated catastrophic destruction of life on earth, and proposed a succession of stage-defining faunas punctuated by extinction events.

Though this catastrophist view was considered eccentric by many of his contemporaries the modern preoccupation with mass extinctions may help to revive interest in his work, even though it might appear crude and simplistic in the light of more sophisticated research undertaken subsequently by Oppel and others.[15]

In this century biostratigraphic correlation and subdivision has become ever more refined and comprehensive, with emphasis being shifted to microfossils and, to a more limited extent, quantitative methods. The use of fossils for correlation has been supplemented by physical and chemical methods, such as magnetic, seismic and isotope stratigraphy, but fossils remain the yardstick by which other methods are assessed. Even today, as interpretive models of environmental change through space and time come and go, achieving a good, enduring biostratigraphic correlation of strata remains one of the most useful things a geologist can do.

Notes

1. Rachel Laudan has given an excellent account of the development of historical geology associated with the Wernerian school in her book (1987), *From mineralogy to geology: the foundations of a science, 1650–1830*. Chicago University Press.
2. Cuvier, G. and Brongniart, A. (1811). *Essai sur la geographie mineralogique des environs de Paris avec une carte geognostique, et des coupes de terrain*. Bandouin, Paris.
3. Laudan, R. (1976). *Centaurus* **20**, 210. For a more conventional treatment of Smith see the chapter by J. M. Hancock in Kauffman, E. G. and Hazel, J. E. (eds), *Concepts and methods of biostratigraphy*. Dowden, Hutchinson & Ross, Stroudsburg, Pennsylvania, 1977.
4. For an absorbing account of Buckland's stratigraphic and palaeoecological work, see N. Rupke (1983), *The great chain of history*. Clarendon Press, Oxford. Rupke argues that Buckland was the effective founder of a distinctive Oxford school of geology, dominated by the stratigraphic-historical approach in contrast to Cambridge, where a mathematical-physical approach was more evident.
5. He persuaded a pet tortoise to walk over the moistened surface of his kitchen table because he thought that a tortoise might have made the traces (N. Rupke, *op. cit.* Note 4).
6. A very thorough account of the controversy is given by J. A. Secord (1986). *Controversy in Victorian geology: the Cambrian-Silurian dispute*. Princeton University Press.
7. Rudwick, M.J.S. (1985). *The great Devonian controversy*. Chicago University Press. Rudwick's account, adopting a strict narrative approach, is even more detailed than Secord's, and establishes a model for scholarship in the history of geology that will be difficult to emulate.

8. de la Beche, H.T. (1839). *Report on the geology of Cornwall, Devon and West Somerset*. Longman, London. Murchison, R.I. (1839). *The Silurian System*. Murray, London.
9. Sedgwick, A. and Murchison, R.I. (1839). *Philos, Mag.* **14**, 241.
10. Sedgwick, A. (1852). *Quart. J. Geol. Soc.* **8**, p. 164.
11. Barrande, J. (1852–1911). *Systeme Silurien du Centre de Boheme*. 8 vols. The author, Prague.
12. Lapworth C. (1879). *Geol. Mag.* n.s. **6**, p. 15.
13. Gressly, A. (1938). *Nouv. Mem. Soc. Helvet. Sci. Nat.* (Neuchâtel) **2** (paper 6).
14. Phillips, J. (1841). *Figures and descriptions of the Palaeozoic fossils of Cornwall, Devon and West Somerset*. Longman, London.
15. For an account of d'Orbigny's work and subsequent developments in biostratigraphic correlation, see chapter by J.M. Hancock in Kauffman and Hazel (*op. cit.* Note 3).

4

The Ice Age[1]

What are nowadays universally seen as deposits directly or indirectly due to glacial activity provided in the early part of the nineteenth century some of the strongest evidence in support of the diluvialist school of thought. Vast tracts of land in Northern Europe and around the Alps had long been known to be mantled by a chaotic assemblage of sediments ranging from muds, silts, sands, and gravels, which frequently had contorted structures, to extremely poorly sorted deposits we now refer to as till, with pebbles and boulders of rock embedded in a fine-grained matrix. A simple bedding sequence in such deposits was frequently impossible to discern. Strangest of all were the large erratic boulders, up to many tons in weight, of evidently far-travelled rock which were found in hill country, plateau, and plain alike, sometimes perched in prominent positions on hillslopes.

Resting upon these deposits in valleys was the more regularly bedded and generally finer-grained river alluvium containing fossils of recognizably modern type. This contrast led Buckland, Sedgwick, and others to distinguish younger *alluvial* from older *diluvial* deposits, the latter having been brought into place by an agency or agencies not operating at the present day, and often containing strange, unfamiliar fossils. As the term 'diluvial' indicated, this agency was generally considered to be a great flood or floods that 'swept over every part of England ... put in motion by no power of nature with which we are acquainted'.[2]

As we have seen, this event was widely identified in the 1820s with the biblical Deluge. The more enthusiastic diluvialists made the most extravagant claims for the geological effects of the flood, which was postulated by Buckland to have submerged the highest summits. It was believed that it was responsible even for what are now interpreted as glacial upland phenomena, such as corries, U-shaped valleys, and polished and striated rock surfaces.

Whereas the naïve correlation of the inferred deluge with the Genesis account was rapidly abandoned after the 1820s, as epitomized by Sedgwick's recantation, belief in the former existence of extremely violent floods sweeping over the earth, with depths of inundation of up to 1500 metres and enormous velocities, persisted for another decade or more in both Europe and North America. Elegant theoretical models were developed of so-called waves of translation, whereby the percussive effects of sudden crustal upheavals raised gigantic ocean waves. Others had speculated more wildly about water spouting from submarine caverns or condensing out of a primordial atmosphere.

Lyell could not tolerate such non-uniformitarian extravagance, whereby for instance huge erratics could be transported by marine currents for hundreds of miles. Yet the various 'glacial' phenomena clearly demanded some unusual explanation. Note was taken of the fact that the bones of reindeer and Arctic birds had been found in deposits as far south as southern France, clearly indicating former episodes of extremely cold climate. Polar exploration had begun to reveal relevant facts, how for instance icebergs were produced by calving from glacier snouts where they entered the sea, and could thence drift for vast distances before eventually melting and thereby shedding their load of unsorted rock debris.

Thus was developed,[3] in moderation of the diluvial theory, the *drift* theory, whereby migrating icebergs grooved underlying rocks and gouged out rock basins. Upon melting far from source, erratics and boulder clay were unceremoniously dumped. The presence of marine shells in some of this glacial 'drift'[4] proved inundation of land by the sea, and there was much lively discussion about the amount of submergence that needed to be invoked for particular areas. The shoreline was invariably drawn at the limit of the drift deposits, and the depth of sea in submerged areas established by reference to the height of striae, erratics, and gravel deposits. Most of Great Britain was thought to have been submerged to a depth of from 300 to 1200 metres and the Alps by more than 2700 metres. Non-catastrophic causes were invoked, such as alternating glaciation and deglaciation of the northern and southern hemispheres, with a corresponding displacement of the earth's centre of gravity.

This was then the setting for the emergence of the glacial theory, which gave a rude shock to the geological community in the late 1830s and early 1840s by invoking the action of formerly far more extensive land ice to account for the varied phenomena that had so intrigued and impressed the diluvialists and adherents to drift theory. Unsurprisingly the source of this revolutionary notion was Switzerland, the only European country with substantial mountain glaciers.

The glacial theory

Naturalists had long been puzzled by the presence of large boulders of granite and schist on the south-eastern flanks of the Jura Mountains, because the bedrock geology of that range was exclusively sedimentary, principally limestone. The resemblance to Alpine bedrock was clear to a number of observers, such as de Luc, who had speculated that the boulders might have been shot violently through the air across the intervening Rhone Valley as a result of localized subsidence compressing air in deep caverns.

This interpretation was scornfully rejected by Saussure,[5] who asked for modern examples of explosions that could shoot large blocks of stone such a great distance. Why did not the blocks smash on impact or bury themselves in the ground? The erratics could always be found *in situ* at the surface. Saussure's careful observations show that the distribution of Alpine moraines could be related to small glacier advances and retreats. The Alpine fragments in the Jura followed the present valley courses, and the places where they had come to rest lay opposite the mouths of the Alpine valleys from which they were derived. Saussure supposed that the erratics had been scattered as a result of a catastrophic *debâcle* involving huge masses of rushing water.

Hutton, who read Saussure avidly in his desire to learn as much as possible about a mountain range he was never able to visit, could not accept de Saussure's explanation, because no flood could have carried erratics uphill for 1000 metres, which is the height of some of them above the Rhone Valley floor. He argued instead that the transport must have taken place *before* the valley was excavated, and goes on to write as follows.

We have but to enlarge our thoughts with regard to things past by attending to what we see at present, and we shall understand many things which to a more contracted view appear to be in nature insulated or to be without a proper cause; such are those great blocks of granite so foreign to the place on which they stand, and so large as to seem to have been transported by some power unnatural to the place from whence they came. We have but to consider the surface of this earth, as having been upon a higher level; as having been everywhere the beds of rivers, which had moved the matter of strata and fragments upon planes, which are, like the haughs of rivers, changing in a continual succession, but changing on a scale too slow to be perceived...

Let us now consider the height of the Alps, in general, to have been much greater than it is at present; and this is a supposition of which we have no reason to suspect the fallacy; for, the wasted summits of those mountains attest to its truth. *There would then have been immense valleys of ice sliding down in all directions towards the lower country,*[6] and carrying large blocks of granite to a great distance where they would be an object of admiration to after ages, conjecturing from whence, or how they came. Such are the great blocks of granite upon the hills of Salève. M. de Saussure, who examined them carefully, gives demonstration of the long time during which they have remained in their present place. The limestone bottom around being dissolved by the rain, while that which serves as the basis of these masses stands high above the rest of the rock, in having been protected from the rain. But no natural operation of the globe can explain the transportation of those bodies of stone, except the changed state of things arising from the degradation of the mountains.[7]

Hutton's closing remarks refer to his belief that the intervening valley must have been eroded before the erratics were transported. Elevation of the Alps would have served two purposes; provide a greater slope towards the Jura and hence increase the transportational power of gravity, and create a colder climate which would provoke the expansion of mountain glaciers.

As usual, Playfair is more lucid and eloquent on the same subject, and his comments lay a greater stress on the geological significance of ice movement.[8]

For the moving of large masses of rock, the most powerful engines without doubt which nature employs are the glaciers, those lakes or rivers of ice which are formed in the highest valleys of the Alps, and other mountains of the first order. These great masses are in perpetual motion, undermined by the influx of heat from the earth, and impelled down the declivities on which they rest by their own enormous weight, together with that of the innumerable fragments of rock with which they are loaded. These fragments they gradually transport to their utmost boundaries, where a formidable wall ascertains the magnitude, and attests the force, of the great engine by which it was erected. The immense quantity and size of the rocks thus transported, have been remarked with astonishment by every observer, and explains sufficiently how fragments of rock may be put in motion, even where there is but little declivity, and where the actual surface of the ground is considerably uneven. In this manner, before the valleys were cut out in the form they now are, and when the mountains were still more elevated, huge fragments of rock may have been carried to a great distance; and it is not wonderful, if these same masses, greatly diminished in size, and reduced to gravel or sand, have reached the shores or even the bottom, of the sea.[9]

One suspects that if such a keen and percipient observer as Hutton had spent some time in Switzerland the glacial theory might have been proposed much earlier than it was. He would surely have appreciated the significance, in the Alpine valleys, of striations on polished surfaces, and roches moutonnées, and relegated to mere folklore the popular interpretations, some of which invoked the action of cartwheels or even hob-nailed boots!

As it was, the prescient remarks of Hutton and Playfair were generally ignored. The indefatigable Buch, for instance, not only confirmed de Saussure's observations on the provenance of the Jura erratics, tracing them back to their source on Mont Blanc and elsewhere, but recognized more generally a radial dispersal of erratics from the Alps. Subsequently he was able to demonstrate conclusively that erratics on the north German and Polish plains must have travelled all the way from Scandinavia.[10] He nevertheless never accepted the Glacial Theory.

In 1815, shortly before Playfair paid his visit to Switzerland, a chamois hunter in Valais, Jean-Pierre Perraudin, came to the conclusion that the glaciers which at that time occupied only the higher portion of the Val de Bagnes, had once filled the whole valley. This was based on the marks or scars he observed on rocks topographically far below the limit of ice, and which matched exactly features to be observed in the rocks surrounding the glaciers.

Within the next three years, Perraudin found a supporter of his view in the civil engineer, Ignace Venetz. Venetz took Perraudin's idea much further. After much detailed observation he realized that moraines occurred far beyond the limits of the present Alpine glaciers, implying that glaciation had formerly been considerably more extensive. In 1821 he felt sufficiently confident to make public his conclusion, at a meeting of a Swiss natural history society. Only slowly, however, did Venetz fully develop his ideas. At the annual meeting in 1829 of the Society of the Hospice of the Great St Bernard, he announced his dramatic conclusion that immense glaciers had once spread out over the Swiss Plain and the Jura Mountains, and indeed other parts of Europe, transporting morainic material and erratics which were dumped upon the eventual melting of the ice.[11]

It appears that Venetz's theory was either ignored or rejected outright at the meeting, but a keen naturalist called Jean de Charpentier, director of the salt mines at Bex, quickly became, after initial scepticism, an enthusiastic supporter. Charpentier spent the next few years making his own detailed field studies, in the course of which he discovered that many uneducated Swiss peasants had long accepted a version of the glacial theory. He encountered such a peasant on the way to Lucerne in 1834 to announce the results of his researches.

Travelling through the valley of Hasli and Lungern, I met on the Brunig road a woodcutter from Meiringen. We talked and walked together for a while. As I was examining a large boulder of Grimsel granite, lying next to the path, he said: 'There are many stones of that kind around here, but they come from far away, from the Grimsel, because they consist of Geisberger [granite] and the mountains of this vicinity are not made of it'.

When I asked him how he thought that these stones had reached their location, he answered without hesitation: 'The Grimsel glacier transported them on both sides of the valley, because that glacier extended in the past as far as the town of Bern, indeed water could not have deposited them at such an elevation above the valley bottom, without filling the lakes'.

In the paper he read at Lucerne[12] Charpentier gave credit due to the pioneering work of Venetz and went on to attribute, for the first time, the extensive polished and striated rock surfaces of the Alps to ice action (previously these had been attributed to the action of gravel-laden water, or something more fantastic). Erratics in the Jura could be traced right back to their source by following the course of Alpine valleys down which they had been transported.

Unfortunately the audience was less sympathetic to de Charpentier than the woodcutter had been, and apparently the general reaction was one of disbelief. One of the sceptics present was the young Swiss naturalist, Louis Agassiz (1807–73), who had been influenced by Cuvier and become an expert on fossil fish. Agassiz had known Charpentier since their schooldays together in Louanne and had a great admiration for him as a naturalist. Nevertheless he was unable to accept his glacial theory initially. In the summer of 1836 he and Charpentier visited the Diablerets glaciers and Rhone Valley moraines. At first Agassiz had hoped to convince his friend of his error, but after some weeks in the area of Bex he himself was converted. Thereafter he proceeded to upstage Charpentier in a way that was to lead eventually to a disruption of their friendship.

Charpentier was a rather shy, mild-mannered, cautious man who lacked the temperament to engage with any enthusiasm in polemical battles with his scientific adversaries, preferring to pursue with great care and scruple his detailed researches. Agassiz was quite different and quickly became an active proselytizer for the cause, developing a comprehensive glacial theory that extrapolated boldly beyond the available evidence in a way that the more conservative Charpentier found difficult to tolerate.

The botanist Karl Schimper was another friend of Agassiz who, following their student days together in Munich, had studied erratics in Bavaria. He lent Agassiz notes indicating that he thought that all Switzerland with most of Europe, northern Asia, and much of North America, had once been covered with thick ice. In February 1837 he wrote a poem to commemorate Galileo's birthday, entitling it *Die Eiszeit: für Freunde gedruckt am Geburtstage Galilei*. This is the earliest reference to an ice age, and the term Eiszeit was quickly adopted by Agassiz, who gave copies of Schimper's poem to those who attended his lectures. This indicates that Agassiz in no way wished to suppress the true parentage of either the term or the concept. Nevertheless, although Agassiz was much more the formidable scientist, who went on to develop the idea in a way probably beyond the capacity of its originator, Schimper became increasingly paranoic in his feeling that he had not received due credit.

Agassiz first presented his glacial theory as a presidential address to the Swiss Society of Natural Sciences of Neuchâtel in July 1837. It consisted substantially of a mixture of the observations and ideas of Charpentier and Schimper. After showing that certain polished and striated surfaces on the slopes of the Jura were exactly like those still being made on glacier floors, Agassiz argued that Alpine ice formerly stretched across the Rhone Valley and rode right over the Jura ranges. More generally, he inferred a general lowering of temperature and added his belief, which went beyond anything inferred by Charpentier, that the ice age had *preceded* catastrophic uplift of the Alps. Agassiz's catastrophist views were also brought to the fore in his account of the mass destruction of animal life by the ice age, most notably Siberian mammoths.

The reaction to Agassiz' lecture[13] was almost entirely adverse, with opponents expressing their opposition in vociferous terms. A participant in the Jura field trip which followed subsequently wrote the following.

In general, I was convinced by my short acquaintance with the leading scientists of the party that a great amount of jealousy and egoism existed between them. Elie de Beaumont was, during the entire trip, as cold as ice. Leopold von Buch was walking straight ahead, eyes on the ground, mumbling against an Englishman who was talking to Elie de Beaumont on the Pyrenees while we were in the Jura, and complaining rather offensively about the stupid remarks made by some amateurs who had joined the group. Agassiz, who was probably still bitter about the sharp criticism made by von Buch of his glacial hypothesis, left the group immediately after departure and was walking a quarter of a league ahead all by himself.[14]

News of the hostile reaction to Agassiz's theory spread rapidly, and later that same year Humboldt urged him by letter to pursue a path of greater caution by returning to the safer field of fossil fish. 'In so doing you will render a greater service to positive geology, than by these general considerations (a little icy besides) on the revolutions of the primitive world, considerations which, as you will know, convince only those who give them birth.'

Agassiz, however, was not a man to be deterred by the sceptics and in 1840 published his monumental memoir on glaciers[15] which, following a lengthy descriptive account of present-day glacial phenomena, outlined the evidence for oscillations of Alpine glaciers in historic time, and concluded with the inference of a vast ice sheet in the past. He thereby anticipated by one year Charpentier's own definitive work on the subject.[16] Charpentier reaffirmed his earlier conclusions about the glacial distribution of erratics, but in contrast to Agassiz, insisted that the period of maximum extension of the glaciers took place *after* the Alps had been uplifted and eroded. In support, he pointed out the way in which transported material occurred in long lines along valley slopes, which could only be explained by assuming that the valleys existed before the glaciers occupied them.

Although Agassiz gave generous acknowledgement to both Venetz's and Charpentier's pioneering efforts, and indeed dedicated his book to them, Charpentier never forgave him for publishing first, and the incident brought their friendship to an end.

One man Agassiz never acknowledged was R. Bernhardi, a professor in the Academy of Forestry at Dreissigacker in northern Germany. Yet Bernhardi anticipated Agassiz by several years by arguing in 1832 that ice from the north polar region had once spread to northern Germany.[17] Bernhardi was acquainted with the work of Venetz and also of Esmark in Norway, who had correctly inferred on the basis of erratic distribution a former much greater extent of Scandinavian ice.[18] It required, however, a lively imagination to bridge the gap between Germany and Arctic ice. Bernhardi's work is all the more remarkable when it is considered that, unlike Venetz and Charpentier, he did not live in close proximity to glaciers. His paper appears, however, to have been totally ignored by his German contemporaries, perhaps because the evidence of erratic distributions in plainlands was in itself much less convincing than the more varied evidence forthcoming from Switzerland.

Agassiz, the glacial evangelist

Agassiz's friendship with Buckland began in 1834, when the latter acted as host in Oxford during part of Agassiz's visit to England to study fossil fish. In reciprocation, Agassiz invited Buckland and his wife to go to Neuchâtel, which they did in 1838.

A study of Buckland's writings reveals that he pondered long and hard over the diluvial and drift theories. Gradual concession over the shortcomings of the former in accounting for erratics did not entirely convince him that the latter was correct. It might be thought therefore that he would be a candidate ripe for conversion to Agassiz's ideas, but he remained cautiously unconvinced by the evidence he saw on a field trip undertaken with his host in the Jura. There followed an Alpine field trip which apparently at last caused a conversion.

Clearly an active campaign of evangelizing was called for, especially
for geologists who had no direct experience of glaciers. The opportunity came in 1840 when Agassiz paid a further visit to Great Britain
with the primary purpose of pursuing his palaeontological studies. He
read a paper outlining his glacial theory to the British Association in
Glasgow. As might be expected the audience reaction was largely
negative, with Lyell leading the critical attack; Buckland remained
silent. After the meeting Buckland persuaded Agassiz and Murchison
to accompany him on a field trip to Scotland and northern England to
examine purported glacial phenomena. This trip finally convinced
Buckland of the essential correctness of the glacial theory and
he subsequently became Agassiz' staunchest supporter in Britain.
Murchison on the other hand was unpersuaded and remained a
supporter of the drift theory to the end.

Buckland soon thought he had persuaded Lyell. In October 1840
he wrote to Agassiz: 'Lyell has adopted your theory *in toto*!!! On my
showing him a beautiful cluster of moraines within two miles of his
father's house, he instantly accepted it, as solving a host of difficulties
which have all his life embarrassed him'. Buckland was over-
optimistic, because Lyell, after initial sympathy with the glacial
theory, returned to his earlier belief in the drift hypothesis.

More important meetings[19] took place later that year at the
Geological Society of London. On 4th November Agassiz read a paper
entitled 'Glaciers, and the evidence of their having once existed in
Scotland, Ireland and England',[20] which described the results of his
trip with Buckland and Murchison. Glacial striae had been recognized at Ballachulish in the Scottish Highlands and on Blackford Hill,
Edinburgh. Roches moutonnées and moraines were found both in the
Scottish Highlands and the English Lake District. The Highlands,
Southern Uplands, Lake District, North Wales, and the Irish Wick-
low Mountains were inferred to be centres of ice radiation, and the so-
called parallel roads of Glen Roy in the Highlands were interpreted as
the successive beaches of a receding glacial lake. 'If the analogy of
facts which he has observed in Scotland, Ireland, and the north of
England, with those in Switzerland, be correct, then it must be
admitted, M. Agassiz says, that not only glaciers once existed in the
British Islands but that large sheets (nappes) of ice covered all the
surface.'

Buckland read a paper at the same meeting. He pointed out that his
attention had first been directed by Agassiz in 1838 to polished,
striated, and furrowed surfaces on the south-eastern slopes of the Jura
near Neuchâtel, and also to the celebrated erratics. It was not,
however, until he had devoted several days to an examination of

actual glaciers in the Alps that he was convinced of the reality of extensive ice action. On his return to Neuchâtel he told Agassiz that he had seen similar phenomena in Great Britain but had hitherto interpreted them as due to diluvial action.

In the discussion that ensued Murchison launched into a critical attack on the glacial theory.

Could scratches and polish just be due to ice? If we apply it to any as the necessary cause, the day will come when we shall apply it to all... If the height be great the result should be proportionate. But in the Highland mountains, not one-third the elevation of the Alps, we have moraines two or three times the magnitude of any known in Switzerland. Formerly, when we have found traces of fragmented rocks disposed around a mountain, we attributed them to the successive periods of elevation in that mountain. The parallel roads of Glen Roy were compared to sea-beaches; now all are attributed to the action of ice.... And if we look to the remains of marine shells found in beds elevated, differing in no respect from those in our present seas, except that they are called 'Pleistocene' [by ... Lyell], we have proof of a lower elevation at the very time ... when these glaciers should be introduced. On these accounts I am still contented to retain our old ideas, that when a mountain was elevated, – or a body of water passed over a series of elevations, the diluvium would descend with the [streams] and be disposed in mounds and terraces according to the direction of the currents etc.

Agassiz replied as follows.

Mr Murchison has objected to the glacial theory in the only way in which it could be objected to. He allows that the whole is granted as soon as you grant a little bit. For here, as in other cases, we argue from what is proved, to what is to be proved. In Switzerland the action of glaciers is yearly seen by thousands of foreigners, and of these facts there can be no doubt, [nor as to the former] extent of glaciers. In the Glacier de l'Aar, grooves, etc. are to be found in the valley twenty two miles from the end of the present glaciers. Did we find these surfaces only on the hard rocks, we might suppose they were merely uncovered by the action of the glaciers; but on the soft limestone rocks these grooves are only to be seen on the surface from which the glacier has just retreated. Many glaciers traverse such rocks only, and there the grooves are annually renewed in winter, and removed by the atmospheric action in summer.... A moraine may be distinguished by certain characters from any other accumulation of fragmented rocks. From the sides of the glaciers moving faster than the middle, there is a continual tendency to throw the fragments into lines at the sides [lateral moraines], and when two glaciers descending from different gorges unite, a medial moraine is formed. The lateral [moraines] are exposed to constant friction with the rocks with which they are brought into contact, and their terminations are passed over by the whole mass of the glacier, so that they become rounded and striated, whilst the medial moraines remaining at the surface remain angular. When the glacier retreats in the summer, the medial moraine, composed of angular fragments, is spread out over the surface of the lateral and terminal moraines, composed of rounded fragments; and it is by these characters that we have proved the existence of moraines in Scotland, Ireland and the north of England...

This cogently argued piece of actualism appears to have made little impression on Greenough, though it should perhaps be borne in mind that this geologist, one of the founding fathers of the Geological Society, had always been sceptical towards abstract theory. Greenough referred to the drastic changes of climate required by the new theory, which seemed very awkward in view of the tropical nature of fossil remains in some recent deposits. Never one to mince words, he considered it to be the 'climax of absurdity in geological opinions. In one period, the Crag, we have three opposite conditions blended: corals, tropical; peat, temperate; shells ... arctic!'

Lyell pointed out in reply that Greenough was confusing distinct episodes of time represented by Coralline, Red/Norwich Crag, and Lacustrine deposits. (The so-called corals of these late Pliocene to early Pleistocene deposits of East Anglia, are in fact bryozoans, with no tropical connotation.)

Buckland summed up 'amidst the cheers of the delighted assembly, who were, by this time, elevated by the hopes of soon getting some tea (it was a quarter to twelve p.m.), and excited by the critical acumen and antiquarian allusions and philological lore poured forth by the learned doctor, who after a lengthened and fearful exposition of the doctrines and discipline of the glacial theory, concluded ... with a look and tone of triumph pronounced upon his opponents who dared to question the orthodoxy of the scratches, and grooves, and polished surfaces of the glacial mountains ... the pains of eternal itch, without the privilege of scratching!'

Two weeks later, on 18 November, another meeting was held, in which Lyell read a paper 'On the geological evidence of the former existence of glaciers in Forfarshire', and Buckland presented the second part of his 'memoir'. In the discussion, Whewell expressed his concern about the necessary decrease in heat from the earth's interior, which has returned since the Ice Age, 'a supposition at variance with all the known laws of physics'. Granted the change of climate, where were the high mountains to act as fulcra for glaciers stretching far into the plain lands?

Buckland stated that he did not explain *all* 'diluvial' phenomena by the operation of glaciers. He allowed the existence of raised bars and beaches, currents, floating and drift ice. But glaciers alone would account for many phenomena in Scotland.

There is no doubt about the stir that Agassiz caused in Britain. In 1841 Edward Forbes wrote to him. 'You have made all the geologists glacier-mad here, and they are turning Great Britain into an ice house. Some amusing and very absurd attempts at opposition to your views have been made by one or two pseudogeologists.'

This was going too far – some of the opposition was quite formidable and there was even a partial relapse among some of Agassiz' supporters, such as Buckland and Darwin. Thus Buckland still continued to apply the mechanism of giant currents as an explanation for some phenomena, which was in fact a recognition of the fluvio-glacial nature of some deposits.[21]

The Cambridge physicist William Hopkins considered that the glacial theory, 'in its application to the transport of blocks across Stainmoor [in northern England] involves such obvious mechanical absurdities that the author considers it totally unworthy of the attention of the [Geological Society.[22] In his presidential address to the Geological Society,[23] Murchison pointed out gleefully that Hopkins' theoretical investigations on the motion of glaciers showed that the locomotion of ice masses over large and flat continents is a theory founded on mechanical error. He thought that the summits of Snowdon in North Wales might have been icy peaks in the midst of an ocean, but that is as far as he would go with the glacial theory. (Hopkins, in his own presidential address a few years later,[24] modified his views to concede that glaciers may indeed act as a transporting power on planes of very small inclination.)

Even in Switzerland, dissenting voices had not been silenced. J.A. de Luc, nephew of the famous naturalist, was a persistent opponent of the glacial theory, as revealed for instance in the following letter to Buckland, written in 1842.[25]

One frequently encounters polished surfaces, striae, furrows on the limestone of the Jura, similar to those produced at the present time at the bottom of existing glaciers. There is no positive proof that glaciers polish and grind the rocks on which they rest, nor that they trace striae and furrows. If a glacier, on melting, leaves a part of its bottom uncovered and there are polished and striated surfaces, that is no reason for believing that it was the ice which produced these effects; they may be very much earlier than the glacier or have been produced by waters which flow constantly under the glacier carrying with them granitic sand and small stones ... The arguments based on observations of the polished and striated rocks in Scotland, rest on the arbitrary supposition that the ice of the glaciers polishes and grinds the rocks.

De Luc claimed to have seen rocks polished by gravel carried along in torrential waters. Thus the arguments of Agassiz as presented, for instance, in his reply to Murchison two years earlier, were unconvincing to at least one of his compatriots.

Far from being subdued by his critics, Agassiz exported the glacial theory to North America after his arrival there in 1846,[26] applying it to the phenomena in the White Mountains of New Hampshire and the Great Lakes. Initially he found only a few converts, but this did not deter him from developing increasingly extravagant views as the years passed, eventually arguing not only for a single ice sheet extending from the Arctic but for an ice-fill even in the Amazon valley!

The triumph of the glacial theory

As might be expected, the drift theory was abandoned soonest in Switzerland, Charpentier's arguments, if not those of Agassiz, proving conclusive to the majority. Other workers such as Guyot were stimulated into tracing erratics back to their source, and finding that the order of succession of Alpine peaks was reflected on the Swiss Plain in parallel lines of erratics. The Chambéry meeting of the Societé Géologique de France in 1844 seems to have marked the final triumph of the glacial theory in Switzerland.[27]

Not that close acquaintance with the Alps was any guarantee of conversion to the glacial theory. Even excursions in the company of Charpentier failed to convince Murchison. In mid-century Murchison went on an excursion with Buch to the Austrian Alps. According to Murchison's report, the aged German geologist ridiculed the modern notion of glacier transport, and halted here and there to plant his staff triumphantly on a big erratic boulder, and energetically demand: 'Where is the glacier that could have transported this block and left it sticking here?'

Though Murchison was also sceptical, as we can appreciate, he was taken aback when Buch began to outline his own ideas. 'When Buch says that the granite blocks on the tops of the Jura were shot across the valley of Geneva like cannon balls by the great power of the explosive force of elevation I feel the impossibility of adhering to him!'[28] Evidently the old man had not abandoned his catastrophic craters of elevation theory, and in this respect at least his ideas on the provenance of the Jura erratics were a throwback to those of the older de Luc. Ironically a granite erratic boulder in Upper Austria, with his name inscribed, serves as a monument to his memory.

In northern Europe, where the evidence was less clear-cut, there was no dramatic abandonment of the drift in favour of the glacial theory. Despite the advocacy of Agassiz, many continued to resist the notion of extensive land ice. A more general view persisted for over a couple of decades that both land ice and submergence had occurred. Thus it was widely believed that widespread glaciation of Great Britain had been followed by a great submergence beneath the sea, in turn succeeded by re-emergence and the ploughing out of moraine drift by valley glaciers. Even Murchison showed a gradual shift of view conceding a greater role to land ice, though it was never regarded by him as having more than local effects. Lyell was unhappy about the glacial theory for a region as far south as Switzerland, and was finally convinced only after a visit in 1857.

Evidently therefore the drift and glacial theories enjoyed an uneasy coexistence for some time in the geological community, with the former being dominant, such that Buckland gave up work on the glacial theory, discouraged by lack of support. The drift theory received, however, a severe blow in 1862 with the publication of a paper by T. F. Jamieson.[29] Jamieson described a dam burst which had caused the sudden emptying of several reservoirs in the Scottish Highlands, and was able to show that the resultant sudden and powerful rush of stone-laden water had not produced either the striae or the deposits associated with glacial action. In some cases sediments had been rapidly transported over surfaces with genuine glacial markings and had failed to obliterate them. The debris eventually deposited by the torrents was quite unlike boulder clay, lacking any mud.

This paper seems finally to have convinced Archibald Geikie, whose own paper[30] published a year later marks, with Jamieson's, a major advance in favour of the glacial theory. Another leading light in the Geological Survey, Andrew Ramsay, wrote an important paper in support at about the same time.[31] Some years earlier Ramsay, carried away with enthusiasm for glacial ideas, had attributed Permian breccias in the English West Midlands, to glaciation. This was based on the large size of the clasts, the angular shape, and the presence on them of polish or striae.[32] This is almost certainly the first instance of a misattribution to glacial action of a pre-Pleistocene rudaceous deposit, because the so-called Clent Breccia has long since been accepted as alluvial fan deposit laid down in a hot desert.

By the mid-1860s Murchison had become a somewhat isolated figure. In his address to the Royal Geographical Society in 1864 he berated the 'ice-mad' geologists such as Ramsay, attacking in particular Ramsay's belief that lake basins had in many cases been excavated by moving land ice and claimed that ice, by itself, lacks the required power to excavate. Ramsay replied by asking why it was that the rivers that flow from glaciers are so muddy. Did it not imply considerable glacial erosion? Murchison evidently failed to appreciate that glaciers consist not only of ice but also of stones.

There remained a general conceptual failure to appreciate what ice sheets were like. Agassiz had conceived of an extensive ice sheet well before a scientific expedition was able, in 1852, to establish clearly that such an ice sheet covered almost the whole of Greenland. Only late in the century was the same thing established for Antarctica. Another reason for the delay in general acceptance of the glacial theory was the extravagance of some of Agassiz' ideas. His notion of a catastrophic uplift of the Alps after glaciation never received general acceptance. Nor did the idea of a single huge ice sheet stretching from the pole survive long. Careful work on the directions of ice movement indicated by striae and erratics established by mid-century that there had been many local ice centres. The recognition that there had not been just one (as Agassiz maintained) but several glaciations was also established quite early, in the 1850s.

Perhaps the most lingering doubt concerned the presence of marine shells in boulder clay, long supposed to be the best evidence of an inundation by the sea. Both Croll[33] and Tiddeman[34] pointed out that such shells in the Scottish and north English 'drift' were as much erratics as the scratched boulders found alongside. They were usually comminuted, and perfect specimens were either small or rare. There was frequently a mixing of species with different temperature and salinity tolerance and quite evidently they had been swept up from the bed of the North Sea and carried along by the moving ice.

The papers of Croll and Tiddeman effectively delivered the *coup de grâce* to the drift theory in Great Britain. In Germany resistance to the glacial theory seems to have been finally eliminated in 1875 when the Norwegian Otto Torell gave an address in Berlin to the Deutsche Geologische Gesellschaft; in this address Bernhardi's stillborn contribution was granted a belated recognition. In North America the glacial theory progressed slowly in acceptance against stubborn resistance, principally through the eloquent advocacy of J. D. Dana, and by 1867 it was almost universally accepted. Conversion was never total, however, the last opposition at the turn of the century effectively dying with the individuals who most strongly expressed it, such as G. M. Dawson in North America and Sir Henry Howarth in England.

Conclusion

One of the most interesting things that emerges from this outline of the long drawn-out battle over the glacial theory is the slow conversion of hypothesis to fact. One cannot conceive of any living geologist who does not accept as established fact that on several occasions in the fairly recent past ice sheets extended over vast areas of Europe and North America now enjoying a temperate climate. Yet less than one and a half centuries ago such a universal acceptance would have been inconceivable, and the subject was one of hot dispute. Strictly speaking, the matter is one of highly probable inference overwhelmingly supported by a wide array of evidence gathered over many decades. Nevertheless modern geologists' assumption that it is purely a factual matter has proved heuristically very valuable in, for instance, the interpretation of sediment cores from the deep ocean. Those breezily pragmatic individuals who insist that 'facts are facts' would do well to ponder on this.

Controversy in this century has centered around the causes of the drastic climatic oscillations in the Pleistocene which have led to repeated glaciations and deglaciations on land. The Serbian mathematical physicist Milutin Milankovitch has pursued at length ideas put forward last century by Croll that a combination of astronomical variables has controlled these oscillations. Milankovitch's thoroughly worked out model based on cyclic changes in the earth's orbital eccentricity, tilt, and precession has in recent years received some support from data obtained from deep-sea cores, but controversy continues, especially about the way such variables can influence climate.[35]

As a postscript, if only because it provides a link with chapter two, we should briefly record Darwin's involvement in the ice age controversy, his interpretation of the parallel roads of Glen Roy, which he came to regard in later life as a great failure of which he was deeply ashamed.

The so-called parallel roads, in the Scottish Highlands, were generally agreed to be natural terraces that signified former beach levels. Previous explanations invoked a former lake that had become dammed at successively lower levels. Darwin dismissed such interpretations in favour of a marine hypothesis, with the beaches being former strandlines elevated as that part of the Highlands rose. No marine shells had been found in Glen Roy, but Darwin argued by analogy with similar terraces in Chile, the 'parallel roads of Coquimbo', which were covered with numerous marine shells.[36] This was good Lyellian reasoning but unfortunately on this occasion the conclusions were erroneous. Agassiz revived the older hypothesis by invoking an ice-dammed lake, with the lower terraces being formed successively as the ice wasted away and the water escaped through newly-emergent cols. Agassiz' interpretation in due course became generally accepted and Darwin himself was converted a quarter of a century afterwards.[37]

Notes

1. The most fully documented account of the controversy is by Charlesworth, J. K. (1957). *The Quarternary era*, Vol. 2. Arnold, London. Much more readable accounts are by North, F. J. (1943). *Proc. Geol. Ass. Lond.* **54**, 1; and Imbrie, J. and Imbrie, K. P. (1979). *Ice ages: solving the mystery.* Macmillan, London.
2. Sedgwick, A. (1825). *Ann. Philos.* **9**, 241.
3. Notably by Lyell, e.g. Lyell, C. (1835). Phil. Trans. Roy. Soc, **1**.
4. The term has been loosely used for glacial deposits up to the present day, although the theory has been long dead.
5. Saussure, H. B. de (1779). *Voyage dans les Alpes*, Vol. 1, p. 150. Barde, Manget, Geneva.
6. My italics.
7. Hutton, J. (1795). *Theory of the earth* Vol. 2, p. 212.
8. Unlike Hutton, Playfair visited Switzerland shortly before his death, and saw for himself the celebrated Jura erratics.
9. Playfair, J. (1802), *Illustrations of the Huttonian Theory*, sec. 349.
10. von Buch, L. (1810), *Reise durch Norwegen und Lappland*. Berlin.
11. Venetz's paper was published in vol. 1 of the *Mem. Soc. Helv. Sci. Nat.* (1833).
12. Charpentier, J. de (1835). *Ann. Mines*, (Paris) **8**, 20.
13. Published as a *Discours* in *Verh. Schweiz. naturf. Ges.* **22**, v (1838). See Corozzi, A. V. (Ed.) (1967). *Studies on glaciers preceded by the discourse of Neuchâtel by Louis Agassiz.* Hafner, New York.
14. Imbrie and Imbrie, *op. cit.* (Note 1).
15. Agassiz, L. (1840). *etudes sur les glaciers.* Neuchâtel. English translation by Carozzi (see Note 13).
16. Charpentier, J. de (1841). *Essai sur les glaciers.* Lausanne.
17. Bernhardi, R. (1832). *Jb. Min. Geognos. Petrefact.* **3**, 257.
18. Esmark, J. (1824), *Mag. Naturvid.* **3**, 28.
19. For a graphic account of these historical occasions see Woodward, H. B. (1908). *The history of the Geological Society of London*, p. 137 *et seq.* Longmans, London.
20. Agassiz, L. (1840). *Proc. Geol. Soc.* **3**, 327.
21. Buckland (1841). *Proc. Geol. Soc.* **3**, 469, 579.
22. Hopkins, W. (1841). *Proc. Geol. Soc.* **3**, 672; (1862) **4**, 90.
23. Murchison (1843). *Proc. Geol. Soc.* **4**, 93.
24. Hopkins, W. (1852). *Quart. J. Geol. Soc.* **8**, xxiv.
25. Cited by North, *op. cit.* (Note 1).
26. He became a professor at Harvard in 1847 and stayed there for the rest of his life.
27. (1844). *Bull. Soc. Geol. France* **2**, 630.
28. Geikie, A. (1875). *Life of Sir Roderick I. Murchison*, Vol. 2, p. 79. Murray, London.
29. Jamieson, T. F. (1862). *Quart. J. Geol. Soc.* **18**, 164.

30. Geikie, A. (1863). *Trans. Geol. Soc. Glasgow* **1**, 145.
31. Ramsay, A. C. (1862). *Quart. J. Geol. Soc.* **18**, 185.
32. Ramsay, A. C. (1855). *Quart. J. Geol. Soc.* **11**, 185.
33. Croll, J. (1870). *Geol. Mag.* **7**, 271.
34. Tiddeman, R. H. (1872). *Quart. J. Geol. Soc.* **28**, 471.
35. Imbrie and Imbrie, *op. cit.* (Note 1) give an excellent account of this twentieth-century controversy.
36. Darwin, C. (1839). *Phil. Trans. R. Soc. London* **39**.
37. Rudwick has give a full account of this subject in Rudwick, M. J. S. (1974). *Stud. Hist. Phil. Sci.* **5**, 97.

5

The age of the earth

It is customary for us nowadays to chuckle at Archbishop Ussher's famous calculation that the earth was created in 4004 BC on 26 October at 9.00 a.m! Such an attitude fails to do justice to the highly scholarly tradition of textual and historical criticism within which Ussher worked, as a figure much respected by his contemporaries. This tradition depended on the assumption that the existence of man was effectively coeval with that of the earth. Consequently it appeared reasonable to believe that the most reliable and accurate way of establishing a chronology of human and earth history was to undertake a critical study of ancient documents, especially calendar systems. This was highly skilled work involving a knowledge of languages, history, and astronomy. Dating by means of calendar systems does of course allow for precision of a day or even an hour. Newton was among the great minds of the period who took part in this kind of chronology and so there is nothing ridiculous about Ussher's calculation when viewed in this context of belief.

More generally, however, there was little concept of the passage of time in the study either of human society or Nature. Such a concept emerged only gradually through the course of the eighteenth century.[1] The seventeenth-century Italian scholar Giambattista Vico was a pioneer in the modern-style historical analysis of human societies,[2] while there developed an increasing sense of recognition that rock strata recorded the passage of time.

The increasing sense that Mosaic chronology was under threat is expressed by William Cowper in his poem *The task*, written in 1785.

... Some drill and bore
the solid earth, and from the strata there
Extract a register, by which we learn
That he who made it, and reveal'd its date
To Moses, was mistaken in its age

Other challenges to the conventional belief that the earth had only a modest antiquity came from the new physical interpretation of the solar system based on Newtonian gravitational theory, notably the independent formulation of the nebular hypothesis by Kant and Laplace, while Buffon calculated an age for the earth of 75 000 years, utilizing the evidence of internal heat and an estimate of the rate of cooling from a presumed former molten mass.

Traditionally, the process of history was seen as one of decline from a paradisal state of perfection to an imminent dissolution of the world to make way for a new heaven and a new earth.[3] The Enlightenment belief in a globe of considerable antiquity became mixed up in the minds of the pious with political events in France, and a reaction grew up in the British Isles in the closing years of the century against the ungodly influences emanating from the continent. This is the proper context in which to view Kirwan's vehement attack on Hutton, for his heretical proposal of what was readily mistaken for an Aristotelian eternalism.

One did not need to be a Huttonian, however, to challenge conventional belief, because the work of Cuvier and other 'catastrophists' established that, very long before the appearance of Man, the earth itself had passed through a process of history consisting of a sequence of 'previous' worlds, each populated by different, unfamiliar, and now-extinct organisms. There also began to emerge a picture of progression through time of various life forms, posing a further challenge to Christian orthodoxy.

This new outlook on time was nothing less than revolutionary, and was common ground among the geologists of the early nineteenth century, but differences began to emerge with the publications of Scrope and Lyell. Both had pressed the need for abundant time to be allowed for the formation of the erosional and depositional phenomena (strata) they had studied. With the triumph of a significant part of Lyell's doctrine by the mid-nineteenth century a widespread belief had become established that virtually unlimited time was at the disposal of geologists.

Whereas Lyell himself was too cautious to make actual estimates of the age of particular geological events, one of his most enthusiastic supporters, Charles Darwin, was less inhibited. In his great book on evolution which appeared in 1859, he stated: 'He who can read Sir Charles Lyell's grand work on the Principles of Geology and yet does not admit how incomprehensibly vast have been the past periods of time, may at once close this volume'.[4]

Darwin went on to illustrate this with reference to the denudation of the Weald in south east England. He compared the total volume of material which must have been eroded from the structural dome with a rough estimate of the rate at which marine denudation would have removed it and arrived at an age of approximately 300 million years. This simple calculation, never intended as more than a rough estimate to illustrate a point, was to mark a turning point in attitudes towards geological time and Darwin was soon to regret his rashness.

Darwin's calculation promptly spurred John Phillips to make an age estimate of his own. Professor of Geology at Oxford and nephew of William Smith, Phillips not unnaturally laid great emphasis on stratigraphy. He questioned the validity of Hutton's 'no trace of a beginning', and was sceptical both about Lyell's doctrine and Darwin's theory of evolution, as indicated in a book that appeared in 1860.[5]

In this book Phillips adopted the cumulative thickness of strata as the best available measure of geological time. Though not a uniformitarian he accepted the practical necessity of assuming uniformity of depositional rate of sediment for his calculation. Relating a total stratal thickness to the average depositional rate should provide a direct measure of the age of the earth's crust. Phillips acknowledged that the calculation requires some additional assumptions, together with estimates of the rate of erosion of existing formations and the relationship of area of land eroded to provide sediment to the area of sea floor to receive it; a 1 : 1 ratio was assumed for the latter. With regard to the former, the Ganges Basin was selected as a region for which relevant data were available.

Phillips assumed that the beds of the Ganges Basin had been deposited at the same rate at which sediment was currently being eroded, and arrived at an age estimate of nearly 96 million years for the formation of the earth's crust. He did not consider this estimate to be other than very approximate. The Ganges carries an unusually heavy sedimentary load, so that any calculation would probably underestimate the age. On the other hand it seemed very probable that the area of sediment accumulation was probably smaller than the area of denudation, so that the result might be too high. Errors could obviously creep into determination of total stratal thickness established by field mapping and stratigraphic logging, applying the principles put forward by his uncle, and the uniformitarian assumption was dubious. Nevertheless Phillips's calculation strongly challenged the Lyellian notion of virtually unlimited time.

The arguments and age estimates of Kelvin

Another person who, for different reasons, reacted adversely to Darwin's age estimate was the great Scottish physicist William Thomson (1824–1907), better known as Lord Kelvin, though he was not elevated to the peerage until 1892 (he will hereafter be referred to simply as Kelvin).[6] Upon asking Phillips's opinion on the subject, the latter replied in June 1861 that 'Darwin's computations are something absurd'.[7]

By the middle of the century, in the years following his appointment at the tender age of twenty-two to the Chair of Natural Philosophy in the University of Glasgow, Kelvin had acquired a considerable reputation as one of the formulators of the basic principles of thermodynamics, in particular the second law. Even before this time he had developed an interest in the relationship between the earth's age and internal temperature and his first published work in 1841 had demonstrated his mastery of Fourier's theory of heat conduction, which was to provide him with his key mathematical tool.

Kelvin was in many respects the archetypal Victorian scientist, of high intellectual ability, wide ranging and successful in his interests, which included the practical applications of science, and with a devout outlook expressed in his belief in the existence of design or divine order. (For this reason he was unable to accept Darwin's theory of evolution by natural selection.) He was indeed widely regarded as the greatest physicist of his time.[8] Furthermore, 'His personality was as remarkable as his scientific achievements; his genius and enthusiasm dominated any scientific discussion at which he was present. He was, I think, at his best at the meeting of Section A (Mathematics and Physics) of the British Association . . . He made the meeting swing from the start to finish, stimulating and encouraging, as no one else did, the yonger men who crowded to hear him. Never had science a more enthusiastic, stimulating or indefatigable leader.'[9]

Not surprisingly therefore, Kelvin was to prove a formidable adversary who dominated thought on the age of the earth during the last four decades of the century, introducing a new approach to the subject backed up with the high prestige of physics.

The first article Kelvin wrote on the subject was in 1862, in a popular magazine, and was entitled 'On the age of the sun's heat'.[10] As in his later work he assumed that all the energy in the universe is gravitational in origin and is constantly being dissipated according to the second law of thermodynamics. In the article he accepted as a high probability that the sun is at present in incandescent liquid body losing heat. Since a finite sun can have neither an infinite source of primitive energy nor a constant mechanical or chemical source of supply, it cannot have existed as a constant source of radiant energy for an infinite period of time.

The sun must, therefore, either have been created as an active source of heat at some time of not immeasurable antiquity, by an overruling decree; or the heat which he has already radiated away, and that which he still possesses, must have been acquired by a natural process, following permanently established laws. Without pronouncing the former supposition to be essentially incredible, we may safely say that it is in the highest degree improbable, if we can show the latter to be not contradictory to known physical laws. And we do show this and more, by merely pointing to certain actions, going on before us at present, which, if sufficiently abundant at some past time, must have given the sun heat enough to account for all we know of his past radiation and present temperature.[11]

This is a very fundamental statement of Kelvin's position to which he was to adhere rigidly in the succeeding decades, in which he stresses the supremacy of 'known physical laws'.

Kelvin singled out the activity of meteors as an existing cause that could be extrapolated back in time. The heat was gravitational in origin, the only other possible source, chemical, being dismissed as totally inadequate. The original potential energy of meteors is progressively converted to kinetic energy on their mutual approach and into heat upon collision. According to the principle of conservation of energy, the heat thereby generated must be exactly equivalent to the original potential energy. Thus if the original aggregate mass of meteors is known and if it is assumed that they were initially at rest relative to each other, then the amount of heat generated by collision can readily be calculated from the known mechanical equivalent of heat. A rapid collision event would give rise to a hot molten mass.

The origins of the idea go back to the Kant–Laplace nebular hypothesis, with the concept of energy being introduced by Helmholz and Kelvin. Kelvin estimated that approximately half the original potential energy of the system would have been dissipated almost immediately but on the other hand the sun's density almost certainly increased considerably with depth, which would allow for more heat generation during formation. Bearing in mind both possibilities he made calculations leading to the following conclusion.

It seems, therefore, on the whole most probable that the sun has not illuminated the earth for 100,000,000 years, and almost certain that he has not done so for 500,000,000 years. As for the future, we may say, with equal certainty, that inhabitants of the earth cannot continue to enjoy the light and heat essential to their life, for many million years longer, unless sources now unknown to us are prepared in the great storehouse of creation.[12]

In this early paper there was little of the dogmatism that was to creep in later, and he was quite cautious in his conclusions, being well aware of the inexactitude of the quantitative estimates. He felt sufficiently confident, however, to give short shrift to Darwin.

What then are we to think of such geological estimates as 300,000,000 years for the 'denudation of the Weald?' Whether it is more probable that the physical conditions of the sun's matter differ 1000 times more than dynamics compel us to suppose they differ from those of matter in our laboratories; or that a stormy sea, with possibly channel tides of extreme violence, should encroach on a chalk cliff, 1000 times more rapidly than Mr. Darwin's estimate of one inch per century?[13]

The implication is clear. If we are forced into a stark choice between a physicist's estimate based on general laws and a geologist's estimate with all the attendant uncertainties about rates of processes, we should opt for the former.

Kelvin dealt more directly with the earth and its secular cooling in another article published a year later.[14] He had long been convinced that the distribution of heat within the earth was the key to its age, and indeed the notion that the earth was originally a hot, molten ball that has cooled gradually dates back to Descartes and Leibnitz. Buffon's crude age estimate, based on this assumption, has already been noted. Whether the earth had resulted from a large number of meteors impacting on a cold nucleus, or, as Kelvin favoured, two large masses colliding, or in some other way, was irrelevant. The only necessary assumption for an age estimate was an originally molten condition.

Kelvin favoured the idea that, on the assumption that molten rock contracts upon cooling, rock cooling near the earth's surface would sink before solidifying, thus causing the formation of convection currents which would maintain thermal equilibrium throughout the whole earth until the start of solidification, which would take place from the centre outwards. When the crust was eventually formed, the earth would consist of a solid sphere at uniform temperature throughout.

The data required to be able to apply Fourier's mathematics consisted of (a) the internal temperature, (b) the thermal gradient at the surface, and (c) the thermal conductivity of rocks. Of these, the thermal gradient was the best known, being on average about 1 °F/50 ft. Kelvin had made his own measurements of thermal conductivity, and it was only the internal temperature that posed serious difficulties. For this it was necessary to apply the solidification model, and assume that the temperature of the uncooled inner core should be comparable to the normal fusion temperature of rocks at the earth's surface.

Kelvin did not play down the difficulties involved in arguing from such evidence and admitted ignorance about the effects of high temperatures in altering the conductivity and specific heat of rocks, and about their latent heat of fusion. Nevertheless he persuaded himself that the data were good enough to provide plausible limits to the earth's age. His best estimate was 98 million years, with a lower and upper possible limit of 20 and 400 million years.

The article dealt explicitly with the Lyellian model of a steady-state earth which was likened to a perpetual motion machine, and it was stressed that such a model took no account of the principles of thermodynamics, notably the second law. While Kelvin did not accept great geological catastrophes he pointed out that the earth's cooling *must* have affected the rates of such processes as volcanic activity, evaporation, precipitation, and erosion, all of which would necessarily have diminished through time.

We must note parenthetically that the Cambridge physicist Hopkins anticipated Kelvin in this respect in a presidential address to the Geological Society of London a couple of decades earlier. He observed that the geothermal gradient suggests a cooling earth and that a cooling earth implied some form of progression through time, which contradicted the uniformitarian doctrine.[15] It corresponded, however, to the beliefs of an earlier generation of geologists, such as Buckland, Sedgwick, and Conybeare.

Kelvin addressed the Geological Society of Glasgow in 1868 and his lecture, on geological time, was published three years later.[16] In this he pursued his attack on uniformitarianism but, for reasons which are by no means clear, he chose to attack Playfair rather than Lyell. (There is in fact no evidence that he ever paid detailed attention to Lyell's writings.)

Most of the lecture was devoted to a third argument which he thought imposed a limit on the earth's age, based on the phenomenon of tidal friction. The tides must act as a kind of brake to slow down the earth's rotation. If one could determine by how much this decelerated annually one could extrapolate back to find the time at which the earth solidified. Kelvin thought he could impose a broad upper limit to the earth's age by this means and, while reluctant to specify a definite figure, believed that the method was at least in principle sound enough to dispose of the uniformitarian argument.

Kelvin spelled out his anti-uniformitarian views most clearly in an article published a few years earlier, which was devoted specifically to the subject.[17]

The 'Doctrine of Uniformity' in Geology, as held by many of the most eminent of British geologists, assumes that the earth's surface and upper crust have been nearly as they are at present in temperature, and other physical quantities, during millions of millions of years. But the heat which we know by observation, to be now conducted out of the earth yearly, is so great that if this action had been going on with any approach to uniformity for 20,000 million years, the amount of heat lost out of the earth would have been about as much as would heat by 100° Cent., a quantity of ordinary surface rock of 100 times the earth's bulk... This would be more than enough to melt a mass of surface rock equal in bulk to the *whole earth*. No hypothesis as to chemical action, internal fluidity, effects of pressure at great depth, or possible character of the substances in the interior of the earth, possessing the smallest vestige of probability, can justify the supposition that the earth's upper crust has remained nearly as it is, while from the whole, or from any part of the earth, so great a quantity of heat has been lost.

The initial reaction

There was little response among geologists to Kelvin's first two papers, which apparently caused him some exasperation, but by the end of the decade he had won an important convert, his friend Archibald Geikie (1835–1924), the newly appointed director of the Geological Survey of Scotland.

In a paper delivered to the Geological Society of Glasgow and published three years later,[18] Geikie surveyed existing knowledge of denudation. The continents are at present being eroded at such a rate that within a few million years they would all be levelled to low plains if no uplift took place. He inferred, without clear evidence, that the present rates of geological processes are probably less than the average through time. Geologists had grossly underestimated the rates of present and past erosional activity and therefore their demands for enormous periods of time were excessive. Geikie upheld the 'new uniformitarianism', of types of process but not their rate or intensity, which could change with time, and uncritically accepted, without independent supporting evidence from geology, Kelvin's estimate for the age of the earth of approximately 100 million years.

Another Scot, James Croll (1821–90), was more critical of Kelvin but also sceptical of the uniformitarians. Croll was a most remarkable man. Of humble origins and self-educated, he was eventually in middle age appointed to a post in the Geological Survey of Scotland and became one of the leading advocates of a limited time-scale and the propounder of an intriguing quantitative theory of ice ages.

In a paper on geological time published in 1868, shortly after he joined the Survey, Croll writes acutely as follows.

But is it the case that geology really requires such enormous periods as is generally supposed? At present, geological estimates of time are little else than mere conjectures. Geological science has hitherto afforded no trust-worthy means of estimating the positive length of geological epochs. Geological phenomena tell us most emphatically that these periods must be long; but how long, these phenomena have, as yet, failed to inform us... They present to the eye, as it were, a sensuous representation of time; the mind thus becomes deeply impressed with a sense of immense duration; and when one under these feelings is called upon to put down in figures what he believes will represent that duration he is very apt to be deceived.[19]

Croll believed that both physics and geology offered good reasons for restricting the earth's age to less than 100 million years. He questioned Kelvin's arguments concerning the sun's heat and completely rejected the argument from tidal friction. Any equatorial bulge that may once have existed must by now have been eroded away and the earth's present shape can give no indication of its past configuration, nor could its assumed rotational deceleration give any indication of age. As regards the sun, Kelvin's hypothesis allowed an age of only about 20 million years, but if one assumed that the particles of the primordial aggregation were not cold but hot, then the time span could easily be increased to about 100 million years. Croll's own method of estimating time, based on his glacial theory, will be dealt with later.

Darwin had been sufficiently shaken by the criticism of the Weald estimate by Phillips at the Geological Society of London and repeated in his book, that in the second edition of the *Origin* he backtracked to the extent of admitting that his figure might need to be reduced by a factor of two or three. For the third edition the calculation was removed, but criticism persisted. Darwin wrote to Lyell; 'Having burned my fingers so consumedly with the Wealden, I am fearful for you ... for heaven's sake take care of your fingers: to burn them severely, as I have done, is very unpleasant'.[20]

Kelvin appears to have been unaware of Darwin's retraction and his criticism left Darwin very disturbed. 'Thomson's views in the recent age of the world have been for some time one of my sorest troubles.'[21]

Darwin sought out Croll's help, because he allowed more time than Kelvin, though still not enough he believed for his evolutionary theory. 'Notwithstanding your excellent remarks on the work which can be effected within a million years, I am greatly troubled at the short duration of the world according to Sir W. Thomson, for I require for my theoretical views a very long period before the Cambrian formation.'[22]

Darwin turned more and more to Lamarckism as a way to speed up the process of evolution and hoped that the 60 million year estimate of the start of the Cambrian could be extended. In the sixth edition of the *Origin* he asked for a suspension of judgement because of our lack of knowledge of the rate of species change in years and also of the earth's interior. He admitted, however, that Kelvin's objection to evolution on grounds of inadequate time was one of the most serious that had been put forward.

The problem was outlined clearly by Fleeming Jenkin in his celebrated review of the *Origin*.

So far as the world is concerned, past ages are far from countless; the ages to come are numbered; no one age has resembled its predecessor, nor will any future time repeat the past. The estimates of geologists must yield before more accurate methods of computation, and these show that our world cannot have been habitable for more than an infinitely insufficient period for the execution of the Darwinian transmutation.[23]

Darwin never came into direct conflict with Kelvin but his self-appointed 'bulldog', T. H. Huxley, took up the challenge with gusto in his presidential address to the Geological Society of London in 1869.[24]

In his 1868 lecture to the Geological Society of Glasgow Kelvin had praised geologists for breaking free of the restrictions of Mosaic chronology, but some geologists had gone too far in their reaction; he singled out Playfair for confusing the *present order* or *present system* with laws now existing. The laws of nature had not changed, but the condition of the earth had. Geological uniformity could not be a law of nature because it contradicted the laws of thermodynamics.

Huxley pointed out in reply that an attack on Hutton and Playfair hardly constituted an attack on modern geological thought.

I do not suppose that, at the present day any geologists would be found to maintain absolute Uniformitarianism, to deny that the rapidity of the rotation of the earth *may* be diminishing, that the sun *may* be waxing dim, or that the earth itself *may* be cooling. Most of us, I suspect, are Gallios, 'who care for none of these things', being of the opinion that, true or fictitious, they have made no practical difference to the earth, during the period of which a record is preserved in stratified deposits.

While these remarks constitute the best defence put forward until then of the modified Lyellian position, Huxley exaggerated somewhat, as befits such a famous polemicist. Both Geikie and Croll had sought a considerable measure of reconciliation with Kelvin, and many if not most contemporary geologists were still probably vague uniformitarians who did not trouble themselves unduly with the restrictions now being imposed on the earth's age. It was sheer sophistry on Huxley's part to criticize Kelvin's large allowances for error as vagueness while at the same time asserting that a factor of two or three would negate his argument. The fundamental concession had, however, been made; Lyell's 'unlimited time' had been abandoned, though it remains open to question just how widely supported this extreme view had ever been.

Turning to the subject of evolution, Huxley was staunchly pragmatic.

Biology takes her time from geology. The only reason we have for believing in the slow rate of the change in living forms is the fact that they persist through a series of deposits which, geology informs us, have taken a long while to make. If the geological clock is wrong, all the naturalist will have to do is to modify his notions of the rapidity of change accordingly.

Huxley then went on the attack; he was the first to point to Kelvin's greatest vulnerability in a way that anticipated the criticisms near the close of the century.

Mathematics may be compared to a mill of exquisite workmanship, which grinds you stuff of any degree of fineness; but, nevertheless, what you get out depends upon what you put in; and as the grandest mill in the world will not extract wheat-flour from peascod, so pages of formulae will not get a definite result out of loose data.

Compare this with the more laconic but decidedly less eloquent modern computer jargon: 'garbage in, garbage out'.

Kelvin replied in equally forceful terms a short while afterwards, and his remarks were published two years later.[25] It was most decidedly not true that few geologists were true uniformitarians. The notion that geologists need not be bothered about the earth and sun cooling down was to be strongly condemned, as was the assertion that limitations of geological time made no great difference to matters biological. 'The limitations of geological periods, imposed by physical science, cannot, of course, disprove the hypothesis of transmutation of species; but it does seem sufficient to disprove the doctrine that transmutation has taken place through "descent with modification".' Geikie and Phillips were cited as geologists who believed that geological and physical evidence could be reconciled. 100 million years was the maximum that Kelvin would allow for the age of the earth, but even 400 million years was insufficient for the uniformitarians.

Whereas Kelvin had some respect for the arguments of geologists, his physicist friend P. G. Tait dismissed them contemptuously and concentrated solely on Kelvin's arguments in making his own calculations on the earth's age, which reduced the time of cooling to at most 15 million years. The revised tidal friction calculation gave a figure of less than 10 million years and the age of the sun was 20 million years at the most.[26]

The reaction to Tait was generally adverse, and his calculations had the inevitable effect of making Kelvin seem a moderate. Accordingly through the succeeding years the younger generation of geologists tended to accept his estimates and endeavoured to adjust their geological thought accordingly; the need to attempt some sort of quantitative age determination had become fairly generally acknowledged.

Independent estimates of geological age

In his 1868 article Croll proposed an entirely original method of geological age determination based on his glacial theory. This adopted the work of the French mathematician Adhémar as a starting point. During periods of high orbital eccentricity the hemisphere experiencing winter at the aphelion of the orbit would suffer long harsh winters whose effects would be only partly offset by the short summers. Consequently ice and snow would accumulate in one hemisphere, while the other hemisphere enjoyed a moderate climate. Because of the precession of the equinoxes these climatic extremes would alternate from one hemisphere to the other every 10 500 years.

Croll went on to calculate the eccentricity for the last million years and found three periods of high eccentricity between 950 000 and 750 000 years together with other less pronounced ones, the two most recent being 200 000 and 100 000 years. He thought that the most recent ice age corresponded to the last period of high eccentricity, which ended 80 000 years ago.

In the tenth edition of his *Principles of geology*, published in 1868, Lyell responded to Croll by rejecting the 100 000 and 200 000 dates as allowing insufficient time for the numerous post-glacial changes to have been accomplished, and adopted 750 000–850 000 years instead for the last ice age. Using his estimate that 95 per cent of fossil molluscs in post-glacial strata are extant, and assuming a time of approximately one million years since the start of the ice age, he inferred a complete 'revolution' (turnover) of species in 20 million years. There had been more than 12 such revolutions since the start of the Cambrian, which could therefore be dated as approximately 240 million years.

Lyell had thus come a long way from his views in the earlier years, and in the eleventh edition (1872) moved even further towards Croll's position by adopting a compromise value of 200 000 years for the end of the last ice age. He refused, however, to accept the shortened time-scale implied, and abandoned any attempt to calculate geological time in years. As for his attempt to counter Kelvin, he found it necessary to invoke the possibility of divine laws at variance with the discovered laws of nature, and hoped that some unknown energy source would eventually be discovered, which indeed happened some three decades later.

Another notable contribution was made by a geological amateur, T. Mellard Reade,[27] who was by profession an architect and engineer. His method of age determination resembled that of Phillips but he differed in his choice of assumptions and parameters. In addition to taking account of mechanical denudation, as Phillips had done, Reade paid considerable attention to chemical denudation, as estimated from soluble minerals occurring in English and Welsh rivers. He also introduced corrections ignored by Phillips, notably concerning volcanic sedimentation and marine denudation.

The principal difference between the calculations of the two men arose from the way in which Reade estimated the volume of sedimentary material in the crust. Starting from the common assumption that the continents and ocean floors have alternately subsided and risen, he concluded that there was no fundamental difference between them. The average depth of both continents and oceans was estimated on somewhat tenuous evidence, to be about ten miles. The assumption was made that the exposed area of the continents undergoing denudation has always been more or less constant and the rates of erosion more or less uniform. Reade then calculated the time necessary for denudation to produce sediments equivalent to a crust ten miles thick over the whole earth and initially came up with a figure of approximately 526 million years. This was treated as a minimum value.

Shortly afterwards Reade undertook another analysis, this time concentrating on limestone erosion and adopting an increased thickness of crust of 14 miles; the new result was 600 million years.

As a consequence of these calculations Reade became a strong opponent of Kelvin, but he was later forced by the results of the *Challenger* expedition to abandon his assumption that the ocean floor is completely covered by land-derived sediment (much of it turned out to be covered by plankton-derived mud). He returned to the Phillips assumption that area of sedimentation is equivalent to area denuded, and also reduced his estimate of crustal thickness, in line with new evidence. The important innovation was introduced of an assumed 'effective area' of denudation (approximately one-third the true area) to compensate for the cyclic reworking of sediments. These various modifications led to a figure of 95 million years for the base of the Cambrian.[28] This is remarkably close to the much older estimate by Phillips based on the cumulative thickness of strata, and in Phillips' time no strata were known older than Cambrian.

Reade had therefore, by the early 1890s, been forced by the

evidence as he saw it into line with Kelvin's earlier age estimate. He happily noted an agreement with Geikie, overlooking the fact that Geikie had himself been influenced by Kelvin.

Samuel Haughton, for 30 years Professor of Geology at Trinity College, Dublin, was an independent-minded thinker who adopted a very original approach to the subject. He worked from the assumption that past changes in climate have been a response to the gradual dissipation of the earth's internal heat, and estimated that 2300 million years had elapsed between the formation of the oceans and the beginning of the Tertiary, a conclusion very acceptable to the uniformitarians.[29]

Thirteen years later he revised his interpretation, though he still adhered to the assumption that climatic changes had followed the secular cooling of the earth.[30] By comparing fossils from the Arctic with analogous forms occurring at the present day in temperate and tropical climates, he constructed a scale correlating past Arctic temperatures with successive geological periods. A ratio of sedimentation rate to stratal thickness was used as a measure of absolute time.

The starting point had necessarily to be when Arctic temperatures descended to 212 °F, at which time water could begin to condense from vapour. Life could only begin when the surface cooled below the coagulation temperature of albumen: 122 °F. According to Haughton's scale based on comparison of fossils, the temperature had reached 68 °F at the start of the Triassic and 48 °F at the start of the Miocene. Since the Miocene the average Arctic temperature had dropped to 32 °F. The amazing conclusion was reached that a greater interval of time had elapsed between the Miocene and the present than between the Triassic and Miocene.

As a result of some strange mathematical juggling Haughton arrived at a figure of only 153 million years for the whole of pre-Miocene time, a drastic reduction on his earlier estimate and without much doubt influenced by a desire to bring his result more or less into line with Kelvin's. Despite the curious assumptions and data manipulations, Haughton's result, like Kelvin's and Croll's, had widespread uncritical acceptance through the 1880s.

Among a variety of other age estimates those of King and Joly warrant special mention. Clarence King, the first director of the United States Geological Survey, adopted a physical approach to the problem, although it was graphical rather than analytical (he was after all a geologist rather than a physicist).

King constructed hypothetical profiles of the earth's internal condition corresponding to the assumed values of his variable parameters and compared these to a similar profile of diabase, a rock which was taken as representative of the earth's crust.[31] An optimum value of

1950 °C was settled upon for the initial temperature and 22–24 million years for the time of cooling. King noted the good agreement with Kelvin's estimate for the age of the sun, though his inferred initial conditions were very different from those of Kelvin.

Needless to say, Kelvin welcomed King's result, but geologists did not, as the tide of opposition rose.

John Joly (1857–1933), who succeeded Haughton as Professor in Dublin, undertook an entirely new approach to the problem near the close of the century, by attempting to estimate the age of the oceans from their sodium content.[32] If it could be assumed that the oceans were initially free of salt, that sodium was supplied at a more or less uniform rate by rivers, and that once it had reached the oceans it remained there, then the simple ratio of sodium in the oceans to rate of supply should give a measure of time. From the data available Joly calculated minimum and maximum values of 90 and 99 million years respectively, depending on different estimates of ocean volume.

Joly analysed in detail a whole variety of complications to arrive at a final result of 80–90 million years, representing the time elapsed since the earth cooled below 212 °F, assuming of course an initial molten condition.

The remarkable agreement with Kelvin was much commented upon, and contemporaries were impressed by Joly's elaborate calculations. Reservations were nevertheless expressed. Thus, while the Oxford geologist W. J. Sollas accepted the method wholeheartedly, he argued for a much greater rate of supply of sodium in the past. Therefore a reduction in the time estimate was required. Osmond Fisher pointed out that the salinity of the oceans back in Silurian times is unlikely to have been much different from today, judging by the marine fossils, and took issue with Joly's reliance on uniformity, with inadequate allowance being made for the recycling of sodium in the erosion–sedimentation cycle. He nevertheless bestowed praise on the novelty of Joly's approach.

At the British Association in 1900, and in a publication of the same year,[33] Joly defended his position with a Kelvinesque assurance and was evidently persuasive. His argument became one of the strongest weapons in the geologists' defence of the 100 million year time-scale, and his influence lasted well into the twentieth century.

The growth of opposition to Kelvin

In the same year that King wrote his paper on the age of the earth, his colleague Upham reviewed the subject of geological time in the same journal.[34]

Reviewing the several results of our several geologic estimates of ratios supplied by Lyell, Dana, Wallace and Davis, we are much impressed and convinced of their approximate truth by their somewhat good agreement among themselves, which seems as close as the nature of the problem would lead us to expect, and by their all coming within the limit of 100,000,000 years which Sir William Thomson estimated upon physical grounds. This limit of probable geologic duration seems therefore fully worthy to take the place of the once almost unlimited assumption of geologists and writers on the evolution of life, that the time at their disposal has been practically infinite. No other more important conclusion in the natural sciences, directly and indirectly modifying our conceptions in a thousand ways, has been reached during this century.

The complacency of those who thought like Upham was to be shattered within a few years, but the groundswell of opposition to Kelvin had begun much earlier.

We recall that in 1868 Croll had been strongly critical of some of Kelvin's arguments, though he agreed on the need to limit geological time. He pursued the matter in his most important work,[35] provoked especially by Tait. The facts of denudation show most convincingly the great antiquity of the earth; it was certain that it must be far more than 20 million years allowed by Tait, though he still believed that 100 million years was long enough. The burden of proof must rest on physicists, not geologists; if there was no agreement then there must be something wrong with the physicists' methods and assumptions.

The criticisms of that remarkable pioneer geophysicist, the Revd Osmond Fisher (1817–1914) were both trenchant and fundamental, though the age of the earth was not his primary concern. He pointed out that Kelvin's calculations concerning cooling of the earth assumed conduction through uniformly solid matter. If, however, the interior were fluid, then convection currents might alter his results appreciably. Fisher wrote an influential book devoted to demonstrating that a plastic substratum supporting a thin crust was essential to account for the character of the earth's surface.[36] Fisher developed a line of attack reminiscent of that first pursued by Huxley, with a more or less explicit charge of arrogance.

With respect to the yielding of the crust, I think we cannot but lament, that mathematical physicists seem to ignore the phenomena upon which our science founds its conclusions, and, instead of seeking for admissible hypotheses the outcome of which, when submitted to calculation, might agree with the facts of geology, they assume one which is suited to the exigencies of some powerful methods of analysis, and having obtained their result, on the strength of it bid bewildered geologists to disbelieve the evidence of their senses.[37]

Geikie was another who had become increasingly repelled by the arrogance of the 'mathematical physicists', despite his early support for Kelvin. In his presidential address to the British Association in 1892[38] he indicated how he had begun to doubt the physical argument.

We must also remember that the geological record constitutes a voluminous body of evidence regarding the earth's history which cannot be ignored, and must be explained in accordance with ascertained natural laws. If the conclusions derived from the most careful study of this record cannot be reconciled with those drawn from physical considerations, it is surely not too much to ask that the latter should also be revised... That there must be some flaw in the physical argument I can, for my own part, hardly doubt, though I do not pretend to be able to say where it is to be found. Some assumption, it seems to me has been made, or some consideration has been left out of sight, which will eventually be seen to vitiate the conclusions, and which when duly taken into account will allow time enough for any reasonable interpretation of the geological record.

Geologists were not the only ones, however, who were beginning to have reservations. G. H. Darwin, who Kelvin had encouraged to pursue the study of tides, cautioned against dogmatism about the earth's age until more precise knowledge was available, in his Presidential Address to section A of the British Association in 1886.[39] He was especially critical of Kelvin's tidal retardation ideas and accused him of attributing to various data a far greater reliability than was justified. He further questioned whether the earth's shape could really give any clue as to its age.

It was left to Kelvin's former assistant John Perry, an accomplished mathematician and engineer, to challenge him directly on his own ground.[40] Perry concentrated upon Kelvin's framework of assumption. If, for example, the earth's thermal conductivity was not homogeneous, as Kelvin assumed, but increased towards the centre, then the age estimate must be increased, perhaps significantly. Like Fisher, Perry insisted on some degree of fluidity, so that the thermal conductivity must be supplemented by convection.

Perry's paper was quickly taken note of and generally welcomed, especially by geologists. It was clearly time for Kelvin to respond to the mounting challenge, which he did in an address in 1897 entitled 'The age of the earth as an abode fitted for life'.[41]

He had lost none of his old assurance; indeed he had abandoned his early caution. After reiterating his older views, starting with a further attack on uniformitarianism and evolution, he dismissed Perry's arguments and suggested, on the basis of new data, that if anything thermal conductivity *diminished* with depth. As for the age of the earth, he was content to accept King's estimate. Thus in his final contribution to the subject the old man had become more entrenched in his views than ever; a possible upper age limit of 400 million years in 1863 had been progressively pared down, to 100 million years in 1868, 50 million years in 1876 and 20–50 million years in 1881, and now only 24 million years in 1897.

Kelvin's address proved the last straw for Geikie. In an explosively polemical reply[42] he effectively declared geologists' independence from the autocratic pronouncements of the likes of Kelvin and Tait. Perhaps the most useful new point made was that evidence on the rates of geological activity seemed to show that they had been remarkably constant since the formation of the oldest known sedimentary strata, whereas according to Kelvin they should have been appreciably higher in the past.

Among those who read Kelvin's address reprinted in *Science* was the distinguished glacial geologist, Professor at the University of Chicago, T. C. Chamberlin (1843–1928). With the astrophysicist F. R. Moulton, he had recently begun to develop an alternative to the nebular hypothesis, namely the so-called planetisimal hypothesis, which implied an origin of the earth by the slow, cold accretion of numerous meteorites. If valid, this hypothesis would completely undermine all Kelvin's arguments. Although the planetisimal hypothesis was not explicitly mentioned, only hinted at, presumably because it was not yet fully developed, it must have influenced Chamberlin in writing the article[43] provoked by reading Kelvin's address.

Chamberlin opened with a courteous acknowledgement of the debt geologists owed to Kelvin.

However inevitable must have been the ultimate recognition of limitations [of time] it remains to be frankly and gratefully acknowledged that the contributions of Lord Kelvin, based on physical data, have been the most powerful influences in hastening and guiding the reaction against the extravagant time postulates of some of the earlier geologists. With little doubt, these contributions have been the most potent agency of the last three decades in restraining the reckless drafts on the bank of time. Geology owes immeasurable obligation to this eminent physicist for the deep interest he has taken in its problems and for the profound impulse which his masterly criticisms have given to broader and sounder modes of inquiry.

Subsequently came the attack. Chamberlin regretted the dogmatism and unjustified assurance implied by such phrases as 'certain truth' and 'half an hour after solidification'. Kelvin's basic assumptions were challengeable. Thus there was no good reason to accept uncritically either the nebular hypothesis or the assumption of a primordial molten earth. The earth could have formed slowly through the gradual accumulation of meteoritic material, rather than through a sudden influx, so that it may never have been molten initially. Indeed, the heterogeneous character of the rocks of the crust was hardly consistent with an initially molten mass. The continued slow influx of meteorites would provide energy to offset the heat lost by radiation. This would extend the permissible time limit.

As for the sun, Chamberlin admitted ignorance, which he invited others to share. One was, however, at liberty to speculate.

Is present knowledge relative to the behaviour of matter under such extraordinary conditions as obtain in the interior of the sun sufficiently exhaustive to warrant the assertion that no unrecognised sources of heat reside there? What the internal constitution of the atoms may be is yet open to question. It is not improbable that they are complex organisations and seats of enormous energies. *Certainly no careful chemist would affirm either that the atoms are really elementary or that there may not be locked up in them energies of the first order of magnitude. . . . Nor would they probably be prepared to affirm or deny that the extraordinary conditions which reside at the center of the sun may not set free a portion of this energy.*[44]

This prescient comment heralds a turning point in the whole controversy; science was on the threshold of a breakthrough in achieving an accurate determination of geological time.

Radioactivity and the geological time-scale

Henri Becquerel discovered the phenomenon of radioactivity in 1896 but the geological significance of the discovery was not recognized until 1903 when Pierre Curie and his assistant found that radium salts constantly release heat. Shortly before this Rutherford and Soddy at McGill University, Montreal, had discovered the enormous energy associated with the release of alpha radiation from any known radioactive material. Rutherford quickly became a leader in the new research field and by 1904 had become convinced that the atoms of all radioactive elements, and perhaps all atoms, contained hugh quantities of latent energy. In the spring of that year he was invited to give a lecture at the Royal Institution in London, and recounted his experience thus.

I came into the room, which was half dark, and presently spotted Lord Kelvin in the audience and realised that I was in for trouble at the last part of the speech dealing with the age of the earth, where my views conflicted with his. To my relief, Kelvin fell fast asleep, but as I came to the important point, I saw the old bird sit up, open an eye and cock a baleful glance at me! Then a sudden inspiration came, and I said Lord Kelvin had limited the age of the earth, *provided no new source of heat was discovered*. That prophetic utterance refers to what we are now considering tonight, radium! Behold, the old boy beamed upon me.[45]

In the course of his lecture Rutherford asserted that the earth could no longer be treated as a cooling body because disintegration of radioactive elements within it liberated enormous amounts of energy. The biological as well as the geological implications, with an increase in the possible time limit for evolution, were duly noted.

Shortly following Rutherford's lecture, Kelvin published a note rejecting the idea that radium could perpetually emit heat, arguing instead that an external source of energy must be invoked. According to Rutherford's biographer,[46] Kelvin publicly abandoned this view at the British Association meeting later that year. This is supported by a comment by J.J. Thomson; Kelvin apparently conceded to him in a conversation that the discovery of radium made some of his assumptions untenable.[47]

Nevertheless Kelvin never actually published a retraction and his last publications indicate an unregenerate attitude. Rutherford's new results were ignored and gravitation was held to be the only possible source of energy. By the middle of the decade, however, most physicists seem to have accepted the notion that radioactivity was in some way responsible for the sun's heat.

As Rutherford and Soddy pursued their researches in a different field, R.J. Strutt began investigating the radioactivity in the earth's crust. His analyses of igneous rocks revealed some 50 to 60 times more radioactivity than the average concentration required to maintain the earth's temperature. Such an anomalous result appeared to be explicable only if radioactive minerals were concentrated in the crust. This discovery proved the final blow to Kelvin's purely mechanical explanation of terrestrial heat.

In 1904 Rutherford made the seminal suggestion that the helium trapped in radioactive minerals might provide a means of determining geological ages. In the following year Boltwood's discovery that lead was invariably associated with uranium ores led him to propose that it might be the stable end product of uranium decay; in 1906 he was convinced, and the implication for age determination was fully appreciated.[48]

The method of age determination would involve determination of the end-product of a radioactive decay series and the time necessary to produce a unit amount of that product from the decay of the parent element. For a given mineral one needed to determine the ratio of the end-product to the parent element. At this early stage in research the available data were very limited, neither the end-products nor the rate of decay being accurately known. Moreover, the necessary experimental work was difficult.

Although Boltwood had by 1907 found a consistently good correlation between the uranium/lead ratios and geological ages of many mineral samples, the latter ranging from 410 to 2200 million years,[49] he did not pursue the matter further. Thus it was Strutt, newly appointed to the chair of physics at Imperial College, London, who became primarily responsible for developing the technique of radioactive dating. Between 1908 and 1910 he calculated the ages of a number of minerals and related his results to their stratigraphic location, using the helium method.[50]

In 1911 a student of Strutt called Arthur Holmes (1890–1965) revived the uranium/lead method pioneered by Boltwood.[51] He subsequently abandoned physics for geology and, more than anyone, was responsible for the development and revision of the geological time-scale in the next few decades.

Holmes' first full review of the various methods of measuring geological time was published in 1913.[52] He noted that both geologists and physicists had to assume for this purpose uniformity in rates of processes of change; both groups could not be correct. The uniform decay of short-lived radioactive elements had been repeatedly demonstrated by experiment and there was no reason to assume that long-lived elements were any different. On the other hand, geological uniformity could not be empirically verified. Furthermore, there was good evidence to suggest that present-day erosional and depositional processes are greater than in the past.

It was, however, a masterly synthesis four years later by the Yale geologist Joseph Barrell (1869–1919) which had really significant impact on the geological community. His article, entitled 'Rhythms and the measurement of geological time',[53] marks a watershed. Before it was published and certainly before Holmes's book appeared, many geologists had, not unreasonably, been sceptical of the new method of radiometric dating, especially when the results of the helium and lead methods did not agree. As Barrell cogently put it,

After one experience with the fallibility of physical argument notwithstanding its mathematical character, it would certainly be unwise for geologists to accept unreservedly the new and larger measurements of time given by radioactivity. There may be here, also, factors undetected and unsuspected which vitiate the results.[54]

After Barrell's paper appeared Joly was almost the only geologist of note who held out strongly against radiometric dating, although he fully appreciated the significance of radioactivity for the thermal history of the earth. He was not content to say that there was something wrong with the new dating method, but set out to prove it. The method was based on the supposition that the radioactive decay constant (λ) had never varied over geological time. The evidence for the constancy of breakdown was, however, only circumstantial. A possibility of testing this arose from the discovery that, in thin sections of certain minerals, such as biotite in granites, there were small circular dark spots or concentric rings called pleochroic haloes, which occurred around tiny inclusions of other minerals, such as sphene and zircon. It was established that these haloes were produced by alpha particles expelled during the decay of radioactive parent elements concentrated in the sphene and zircon. In addition it was known that for a given host mineral and a given radioactive parent, the distance travelled by such alpha particles is proportional to the decay constant (λ). Therefore the size of the haloes from minerals of different ages could be used as a measure of the possible variation of λ with time. Evidence was produced that uranium haloes were larger in Precambrian than in younger rocks, pointing to a decrease in λ with time.

Difficulties of this sort were resolved as further research gradually clarified the whole process of radioactive decay, in particular as a result of the discovery of isotopes. It was recognized that radioactive decay was dependent on certain isotopes with unstable nuclei, and that these ejected particles until a final stable state was reached. Therefore, to achieve accurate results from radiometric dating it was necessary to determine the amounts of both parent and daughter products in terms of isotopes rather than as elements, which was all that had been possible using normal chemical methods of analysis. Such determinations only became possible with the development of the mass spectrometer, which separated atoms from one another according to their mass.

By 1930 it had been shown by the use of this instrument that there were two isotopes of uranium (with masses of 238 and 235) and one of thorium (mass 232) involved in natural radioactive decay, and that each of these isotopes had very different half-lives. Radiometric age could be determined by evaluating the ratio between these three parent isotopes and their daughter products, lead, i.e. $^{238}U/^{206}Pb$, $^{235}U/^{207}Pb$ and $^{232}Th/^{208}Pb$. If, in a particular mineral or rock, the age given by any one of these methods was not consistent with any of the others, the results could not be considered reliable. This greater understanding of the radioactive process seemed to resolve the controversy over the size of pleochroic haloes, for it was now possible to explain the large haloes in Precambrian rocks in terms of the faster decay of ^{235}U and the smaller haloes as a result of the slower decay of ^{238}U. This was sufficient to refute Joly's contention that λ varied with time.

Barrell pointed out that the doctrine of uniformitarianism had ignored the significance of rhythms in nature, but both erosion and sedimentation are pulsatory processes. Sedimentation is dependent on changes in base level. The rate of sedimentation is determined less by the rate of supply than the rate of discontinuous depression of the surface of deposition. Such depression must precede sedimentation, though sediment load will reinforce it. Deposition is not normally a continous process; there are numerous small-scale interruptions, which he termed *diastems*.

Because of the great elevation of the continents at the present day and the magnitude of recent orogenic movements, the present rates of denudation and consequently sedimentation are very much greater than the mean for geological time. It follows from these arguments that most geologists have grossly underestimated the time significance of sedimentary strata. Barrell considered that, on geological grounds alone, 250 million years might be regarded as a reasonable estimate of the time elapsed since the start of the Cambrian. Radiometric dates indicated that this figure needed to be increased by a factor of two.

Barrell's time-scale, like the more refined versions that Holmes put forward in later years, depends firstly on the principle enunciated many years before by Haughton: 'the proper relative measure of geological periods is the maximum thickness of strata formed during those periods.' Secondly, one must accurately date minerals whose relative age is known from the stratigraphical setting. Thus, for the lead method used by both Barrell and Holmes, pitchblende-bearing pegmatites might cut Upper Silurian and be unconformably overlain by Lower Devonian strata. A uranium/lead age determination should therefore indicate approximately the absolute age of the end of the Silurian.

As the years passed the number of reliable dates inevitably increased, as did the variety of isotopes utilized in radiometric dating and which could therefore be used for cross-checking. Instrumental quality and hence precision of age determination improved, and so did the understanding of geological relationships bearing on age determinations, while considerable new information accumulated on stratal thickness across the world. Despite all this there has been no really drastic revision of the Phanerozoic time-scale in over half a century, as Table 2 clearly demonstrates. Yet in the closing years of the last century and the start of this there was a wild disparity in the dates put forward by the geologists who attempted age determinations.[59] Thus the

Table 2   *The Phanerozoic time-scale in millions of years*

Base of	Barrell[53] (1917)	Holmes[55] (1933)	Holmes[56] (1947)	Holmes[57] (1960)	Lambert[58] (1971)
Pleistocene	1–1.5	1	1	1	
Pliocene	7–9	15	12–15	11	7
Miocene	19–23	32	26–32	25	26
Oligocene	35–9	42	37–47	40	38
Paleocene–Eocene	55–65	60	58–68	70	65
Cretaceous	120–50	128	127–40	135	135
Jurassic	155–95	158	152–67	180	200
Triassic	190–240	192	182–96	225	240
Permian	215–80	220	203–20	270	280
Carboniferous	300–70	285	255–75	350	370
Devonian	350–420	350	313–18	400	415
Silurian	390–460	375	350	440	445
Ordovician	480–590	440	430	500	515
Cambrian	550–700	510	510	600	590

base of the Cambrian was variously determined, between 1885 and 1902, at 3, 18, 28, 600, 794, and 2400 million years! Such is the measure of value of Barrell's achievement in particular and radiometric dating in general.

As regards the age of the earth, there was an important meeting of the National Research Council in Washington in 1931. In the longest and most important section of the committee report[60] Holmes concluded: 'No more definite statement can therefore be made at present than that the age of the earth exceeds 1460 million years, is probably not less than 1600 million years, and is probably much less than 3000 million years.' Half a century later it is generally believed that the earth is approximately 4500 million years old, though the age of the oldest rocks yet discovered is less than 3800 million years. There may yet be further revisions, but they are likely to be insignificant. The age of the earth is no longer the subject of major controversy.

Since the age of the earth is the title of this chapter it is desirable to conclude with an outline of the modern method for determining this. The method depends upon accepting meteorites as aggregates of minerals that have been radiogenically closed systems since the time of planetary formation, early in the history of the Solar System. On analysis, iron meteorites were found to contain such extremely small amounts of uranium and thorium that it could be assumed that the lead present was in no way radiogenically derived and hence must have been primevally inherited. As such, this was lead that had remained virtually unchanged since the time of planetary formation. The higher values of uranium and thorium found in stony meteorites meant that their lead was a mixture of both primeval and radiogenic sources. As the primeval lead contribution was known, from the iron meteorite data, the radiogenic contribution could be determined. From the ratio of radiogenic $^{207}Pb/^{206}Pb$, an age for stony meteorites was determined as approximately 4500 million years; iron meteorites can be assumed to have the same age.

As this can be taken as an indication of the time of general planetary formation, it is probable that the earth formed at about the same time. The problem then was to obtain hard evidence to confirm this. The closed systems of the meteorites are thought to be derived from materials equivalent to the mantle and core of the earth. The material that is available from the lithosphere is being profoundly affected by so many processes that no rocks can be found that have maintained a closed system for anywhere near this length of time. The solution lay in considering the earth as a whole as a closed system. In these terms, since the earth was formed, lead from radiogenic sources has been continuously added to the original stock of primeval lead, so that if a good average sample of the earth's lead at the present day could be obtained it would be equivalent to the lead of a stony meteorite. The red clay and manganese nodules of the Pacific Ocean floor contain such a sample of lead, for it has been derived from a wide variety of sources distributed all round the Pacific Basin and thoroughly mixed in the oceanic waters, prior to its deposition and incorporation of these materials. Plotting the ratios of radiogenic and primeval lead $^{207}Pb/^{204}Pb$ against $^{206}Pb/^{204}Pb$, resulting from analyses of these nodules and clay, locates a point on the 4500 million year so-called isochron derived from meteorites. That is, the earth was formed 4500 million years ago at the same time as the meteorites.

An epilogue on the catastrophist–uniformitarian debate

Modern writers on uniformitarianism[60] have agreed that there was some confusion of thought in the doctrine that Lyell expounded. In Gould's terminology, Lyell failed to distinguish clearly between a *methodological uniformitarianism*, comprising the assumption that natural laws are constant in space and time, and a *substantive uniformitarianism*, which postulates a uniformity of material conditions or of rate of processes. The former, which also demands that no hypothetical unknown processes be invoked if observed historical results can be explained by presently observed processes, is vital for the interpretation of past events. It is in effect synonymous with actualism, and was readily conceded by such early opponents as Sedgwick.

Dispute has been essentially confined to the notion of substantive uniformitarianism. It has become almost fashionable to say that Lyell was wrong, but the issue is more complicated.

On the question of organic progression through time Lyell was clearly in error, as he was forced eventually to concede, coming reluctantly to accept Darwin's theory of evolution after several years of opposition. The question of progression in the physical world was rather different. Kelvin's attack would have been devastating only to an Aristotelian theory of eternalism, which Lyell most decidedly never embraced. Whereas his vagueness about time, apart from the fact that he envisaged plenty of it, might be considered a shortcoming, it reduced his vulnerability to Kelvin's onslaught, and it was Darwin who caught the most severe blast for his presumption. It is important to observe that the undermining of Darwin's assumptions and subsequent vast extension of the known age of the earth did not alter his fundamental point about there being some sort of directionality through time, it merely weakened its impact.

The essential point is surely that Lyell's uniformitarian doctrine was a less comprehensive 'theory of the earth', to use a phrase from an earlier era, than a heuristic principle. As such its value was immense. Consider for instance the early conflicts with the catastrophists over two of their principal claims, of paroxysmal uplift of mountain ranges and the decrease in intensity of igneous activity through time. Progressively all but the most diehard opponents were forced to retreat from such claims as a result of a combination of stratigraphic research and the application of Lyellian principles. Huxley was right – Kelvin's work had no direct relevance to most geological research because there was still more than sufficient time in 100 million years to accomplish production of the most spectacular phenomena. As eminent and sensible a geologist as Geikie only balked when his erstwhile ally subsequently wished to pare down the figure to a fraction of this value.

There is a widely accepted modern view of a physical world that has shown no significant directionality of change in rates or types of geological processes, or in the composition of the lithosphere, hydrosphere, and atmosphere, since the early Precambrian. Bearing in mind that the Precambrian was a closed book to nineteenth-century geologists, the Lyellian interpretation that what we nowadays call the Phanerozoic record signifies a steady-state earth seems a remarkably sound first-order approximation.

This does not mean, however, that the catastrophism–uniformitarianism debate is dead, because recent years have seen the emergence of what can be called neo-catastrophist school of thought. While rejecting the cruder notions of the early catastrophists, adherents embrace a punctuated view of geological (and biological) history markedly at variance with that of Lyell and Darwin but corresponding to that of Barrell. Thus the stratigraphic record has been compared to the traditional life of a soldier, long periods of boredom interrupted by moments of terror![61]

Sedimentologists have become increasingly interested in the activities of the rare turbidity current or intense storm, which may have more significant erosional and depositional consequences every decade or so than the modest everyday activities more amenable to direct investigation. Lyellian extrapolation of such processes to larger events, simply adding the time dimension to increase magnitude by a cumulative effect, may often not be justifiable.

There is a parallel shifting of thought among geomorphologists. Thus the once heretical conclusion that the channelled scablands of eastern Washington State, USA, were eroded by catastrophic floods, has in recent years achieved widespread acceptance.[62] On a larger scale, Gilluly's 'null hypothesis' that tectonic activity is random in time if not in space,[63] put forward in opposition to Stille's theory of brief, world-wide orogenic episodes, is now thought to be a considerable overstatement. There is even a new 'catastrophe theory' in mathematics which has found widespread application in engineering studies and elsewhere. It postulates that a gradually changing *cause* may be reflected not in a gradually responding *effect* but in a sudden shift from one stable state to another.[64] This could well be relevant to crustal tectonics, with the slow accumulation of stress being relieved periodically by violent and destructive earthquakes.

The question is just as relevant to organic evolution. Darwin was so wedded to the notion of gradualistic change, presumably derived from his hero Lyell, that he was positively embarrassed by the numerous gaps and lack of transitional forms in the fossil record. His explanation, that this is a consequence of the extreme imperfection of the stratigraphic record, with many intervals of time being unrepresented, has become progressively less plausible as the years have passed. Ever-expanding research activity across the world, including the ocean floor, has filled out the stratigraphic record to an enormous extent, but many of the fossil discontinuities have persisted.

Eldredge and Gould[65] grasped the nettle by proposing that the conventional gradualistic view of species change through time was in error. They put forward an alternative, termed punctuated equilibria, in which long periods of morphological stasis are interrupted by brief, geologically 'instantaneous' episodes of speciation, when significant genetic, and consequently morphological, change takes place. By no means all palaeontologists have yet been persuaded, though there is much empirical evidence in support of this new interpretation.

With regard to whole groups of organisms mass extinction episodes such as those at the close of the Palaeozoic and Mesozoic eras are widely considered to be the result of some global catastrophe, though the precise nature of such catastrophes remains a matter of dispute.[66] (Mass extinctions are the subject of chapter 7.) Similarly, the reverse phenomena of organic radiations are seen as occupying geologically brief episodes compared with the periods when no dramatic changes in diversity, or emergence of major new groups, took place. This is especially true of the dramatic radiation of the Metazoa in the early Cambrian. We have moved a long way from the view of Darwin, who pointed out in his letter to Croll that '... I require for my theoretical views a long period before the Cambrian formation'. Whereas there is of course a record of primitive unicellular life extending back to the early Precambrian, the widespread and long-standing belief in a lengthy Precambrian history of unskeletonized 'higher' organisms is no longer considered tenable.

Notes

1. Toulmin, S. E. and Goodfield, J. (1965). *The discovery of time.* Hutchinson, London.
2. Berlin, I. (1979). *Against the current.* Hogarth Press, London.
3. Porter, R. (1977). *The making of geology.* Cambridge University Press.
4. Darwin, C. (1859). *On the origin of species* (1st edn), p. 282. Murray, London.
5. Phillips, J. (1860). *Life on earth: its origin and succession.* Macmillan, London.
6. An invaluable book for the study of Kelvin's involvement in the age controversy is Burchfield, J. D. (1975). *Lord Kelvin and the age of the earth.* Macmillan, London.
7. Thompson, S. P. (1910). *The life of William Thomson, Baron Kelvin of Largs,* Vol. 1, p. 539. Macmillan, London.
8. This is a judgement that has not survived. His Scottish contemporary J. Clerk Maxwell was a thinker of much greater originality and enormous influence on the physics of a later era.
9. J. J. Thomson (himself a very distinguished physicist) quoted in King, A. G. (1925). *Kelvin the man,* p. 96. Hodder and Stoughton, London.
10. Kelvin, Lord (1862). *Macmillans Magazine* **5**, 288.

11. Kelvin, *op. cit.* (Note 10), p. 370.
12. Kelvin, *op. cit.* (Note 10), p. 375.
13. Kelvin, *op. cit.* (Note 10), p. 368.
14. Kelvin, Lord (1863). *Phil. Mag.* (ser. 4) **25**, 1.
15. Hopkins, W. (1852). *Quart. J. Geol. Soc. Lond.* xxiv.
16. Kelvin, Lord (1871). *Trans. Geol. Soc. Glasgow* **3**, 1.
17. Kelvin, Lord (1866). *Proc. Roy. Soc. Edinb.* **5**, 512.
18. Giekie, A. (1871). *Trans. Geol. Soc. Glasgow* **3**, 153.
19. Croll, J. (1868). *Phil. Mag.* (ser. 4) **35**, 363.
20. In Darwin, F. and Seward, A. C. (eds) (1903), *More letters of Charles Darwin*, Vol. 2, p. 139. Murray, London.
21. C. Darwin, letter to Wallace (April 1869). In Marchant, J. (ed.) (1916). *Alfred Russell Wallace, letters and reminiscences*, Vol. 1, p. 242. Cassell, London.
22. C. Darwin, letter to Croll (January 1869). In Darwin and Seward, *op. cit.* (Note 20), Vol. 1, p. 313.
23. Jenkin, F. (1867). *North British Review* June, p. 277.
24. Huxley, T. H. (1869). *Quart. J. Geol. Soc. Lond.* **25**, xxxviii.
25. Kelvin, Lord (1871). *Trans. Geol. Soc. Glasgow* **3**, 215.
26. Tait, P. G. (1876). *Lectures on some recent advances in physical science.* Macmillan, London.
27. Reade, T. M. (1878). *Proc. Liverp. Geol. Soc.* **3**, 211; Reade, T. M. (1879). *Chemical denudation in relation to geological time.* Bogue, London.
28. Reade, T. M. (1893). *Geol. Mag.* (ser. 3) **10**, 97.
29. Haughton, S. (1865). *Manual of geology*, p. 99. Longman, London.
30. Haughton, S. (1878). *Nature* **18**, 266.
31. King, C. (1893). *Am. J. Sci.* **145**, 1.
32. Joly, J. (1899). *Smithsonian Rep.*, 247.
33. Joly, J. (1900). *Geol. Mag.* (ser. 4) **7**, 124.
34. Upham, W. (1893). *Am. J. Sci.* **145**, 209.
35. Croll, J. (1875). *Climate and time in their geological relations.* Murray, London.
36. Fisher, O. (1881). *Physics of the earth's crust.* Macmillan, London.
37. Fisher, O. (1882). *Geol. Mag.* (ser. 2) **10**, 94.
38. Geikie, A. (1892). *British Assoc. Rep.*, p. 3.
39. Darwin, G. H. (1886). *British Assoc. Rep.*, p. 511.
40. Perry, J. (1895). *Nature* **51**, 224.
41. Kelvin, Lord (1899). *J. Victoria Inst.* **31**, 11.
42. Geikie, A. (1899). *British Assoc. Rep.*, p. 718.
43. Chamberlin, T. C. (1899). *Science* **9**, 889; **10**, 11.
44. Chamberlin, T. C. (1899). *Science* **10**, 12. My italics.
45. Eve, A. S. (1939). *Rutherford*, p. 107. Macmillan, New York.
46. Eve *op. cit.* (Note 45).
47. Thomson, J. J. (1936). *Recollections and reflections*, p. 420. Bell, London.
48. Badash, L. (1968). *Proc. Am. Phil. Soc.* **112**, 157.
49. Boltwood, B. (1907), *Am. J. Sci.* **23**, 77.
50. Strutt, R. J. (1910). *Proc. R. Soc. London* **A84**, 194.

51. Holmes, A. (1911). *Proc. R. Soc. London* **A85**, 248.
52. Holmes, A. (1913). *The age of the earth*. Harper, London and New York.
53. Barrell, J. (1917). *Bull. Geol. Soc. Am.* **28**, 745.
54. Barrell, J. *op. cit.* (Note 53), p. 821.
55. Holmes, A. (1933). *J. Wash. Ac. Sci.* **23**, 169.
56. Holmes, A. (1947). *Trans. Geol. Soc. Glasgow* **21**, 117.
57. Holmes, A. (1960). *Trans. Edinb. Geol. Soc.* **17**, 183.
58. St. J. Lambert, R. (1971). *Geol. Soc. Land. Spec. Publ.* no. 5, p. 10.
59. *Nat. Res. Council Report*, Bull. no. 80 (Washington).
60. e.g. Hooykaas, R. (1963). *The principle of uniformity in geology, biology and theology*. Brill, Leiden. Gould, S. J. (1965). Is uniformitarianism necessary? *Am. J. Sci.* **263**, 223. Hubbert, M. K. (1967). Critique of the principle of uniformity. In *Uniformity and simplicity* (ed. C. C. Albritton). Geol. Soc. Am. Spec. Paper no. 89, p. 3. Simpson, G. G. (1970). Uniformitarianism. An enquiry into principle, theory and method in geohistory and biohistory. In *Essays in evolution and genetics in honour of Theodosius Dobzhanksy* (eds M. K. Hecht and W. C. Steere), p. 43. Appleton, New York.
61. Ager, D. V. (1973). *The nature of the stratigraphic record*. Macmillan, London.
62. Bretz, J. H. (1969). *J. Geol.* **77**, 505.
63. Gilluly, J. (1949). *Bull. Geol. Soc. Am.* **60**, 561.
64. Zeeman, E. C. (1976). *Sci. Amer.* **234**, 65.
65. Eldredge, N. and Gould, S. J. (1972). In *Models in paleobiology* (ed. T. J. M. Schopf), p. 82. Freeman Cooper, San Francisco.
66. Rudwick has pointed out, in *The meaning of fossils*, that Cuvier's theory of catastrophe-imposed extinctions was empirically well supported.

6
Continental drift

Unlike the controversies outlined so far, that concerning continental drift has been confined essentially to this century, although the idea that the continents might have moved laterally with respect to each other had been proposed earlier.

The highly successful modern theory of plate tectonics, which has an all-pervasive influence on the earth sciences today, may be regarded as an outcome and development of the continental-drift hypothesis, which provoked intense controversy for half a century following its presentation by Alfred Wegener. For many years its adherents were often dismissed contemptuously as cranks by the geological and geophysical establishment on both sides of the Atlantic, though more especially in North America. At best the idea was considered inadequately supported by evidence and mechanically implausible, and it had no serious effect on the mainstream interests of most earth scientists. The story of how the consensus was converted is one of the most fascinating and best documented in the history of science.[1]

Before dealing with the controversy it is necessary to outline the conceptual model of the earth that dominated late nineteenth and early twentieth century thought, to appreciate the revolutionary nature of Wegener's hypothesis. It is also desirable to note the contributions of some of Wegener's precursors.

The cooling, contracting earth

Whatever the merits of the *Principles of Geology*, and they are considerable, Charles Lyell failed to provide a satisfactory explanation for one of the most challenging of all geological problems, the origin of mountain ranges. There was a much greater preoccupation with this question in continental Europe than in Britain, starting with Saussure's work in the Alps in the late eighteenth century. The leading figure for much of the nineteenth century was Leonce Elie de Beaumont, a pioneer in the application of mathematical techniques to geological problems. Building on the ideas of Leopold von Buch, the correlation of age and strike of mountain chains became a principal element in his theory, which also sought to relate Cuvier's organic extinction phases to episodes of catastrophic upheaval, as noted in chapter 2. The energy for these upheavals was derived from global contraction due to progressive cooling. The notion of a contracting earth that was initially molten and was now in the process of cooling and solidifying was widely accepted in the nineteenth century, for example by both Kelvin and his geological critics. Elie de Beaumont's decidedly non-Lyellian earth model had enormous influence on the continent, but it became restrictive in later years when he became obsessed with great circle intersections in his reseau pentagonal theory.[2] In the mid-nineteenth century, gravity studies indicated that the Himalayas were apparently exerting much less gravitational attraction than expected from their enormous mass. Before long it became generally accepted that this must be the consequence of the lighter rocks of the mountains projecting in some way down into the underlying crust. G. B. Airy put foward the proposal that beneath the earth's solid crust there is a layer of material that acts in the long term as a fluid, and is denser than the solid crust that can be treated as in effect floating on it. If this crust is thicker in some places, as appears to be the case in mountainous regions, its base would sink into the underlying material until buoyancy provided by this material equalled the mass of the mountain. Conversely topographic depressions such as those of the ocean basins would be underlain by an upward bulge of the denser subsurface material beneath a much thinner crust. If the crust were thinned by erosion or thickened by the accumulation of a pile of sediments, there should be provoked an adjusting uplift or subsidence respectively because of the removal and addition of load.

This principle of *isostasy*, and the notion of an ocean crust composed of basalt, was incorporated in 1873 into a more general model of the earth by the distinguished American geologist J. D. Dana (1813–95).[2] Dana proposed that at the time of the initial solidification of the surface, large areas had a granitic composition whereas others were composed of basaltic crust. The crust would therefore have to accommodate itself to this contraction and this would lead to the development of lateral compressive forces within the crustal zone. Because of the difference in level between the depressions and the plateaus, the basaltic crust of the depressions would act as a lever against the granitic crust. Consequently the lateral pressure would be directed from the oceanic depressions towards the continental plateaus.

At an early stage such pressures would have caused an overall bending or flexing of the plateaus, producing broad swells rising above sea level and equally broad depressions. This would have initiated the erosion and transportation of rock from the raised areas and its deposition as sediments in the depressions. The process was continued by *isostatic adjustment*, the term given to the vertical movement of a section of the earth's crust in response to an increase or decrease in load, consequent on the erosion and deposition, combined with continued lateral pressure. As the sediments in the basins were depressed to greater depths they entered zones of higher temperatures which greatly weakened or even melted them. Lateral pressure could be relieved by intense folding and fracturing within the zone of weakening leading to the formation of mountain belts with folded strata. (This is the origin of the *geosynclinal* concept for which Dana is most remembered today.)

The model therefore explained the formation of continents and ocean basins and, within the continents, the difference between fold mountains, plainlands, and the continental shelf. It was applied with most success to North America, where the most important mountain ranges are adjacent to either the Pacific or Atlantic Oceans.

A comparable earth model was developed by the Austrian geologist Edward Suess (1831–1914), in his great multivolume treatise *Das Antlitz der Erde*, published near the end of the nineteenth century.[3] In the course of progressive solidification and contraction from a molten mass, lighter rock materials had moved towards the surface to give rise to granitic-type igneous and metamorphic rocks, with associated sediments. They were collectively termed *sal* (later changed to *sial*) because they were relatively rich in silicates of alumina, together with soda and potash. Underlying the sial were denser rocks termed *sima*, resembling, if not exactly matching, basalt, gabbro, or peridotite, which are rich in silicates of iron, calcium, and magnesium.

Mountain ranges were produced by contraction in a manner somewhat analogous to the crinkles developed on a shrinking, drying apple. On a larger scale, an overall arching pressure caused certain sectors of the earth's surface to collapse and subside, giving rise to the oceans, while the continents remained emergent as unfaulted blocks or 'horsts'. In the course of time certain continental areas in turn sank faster than adjacent areas and hence were inundated by the sea, while temporarily stabilized parts of the ocean floor at other times emerged once again as dry land.

Evidence of former land connections across what was now deep ocean was provided abundantly by the total or near-identity of many fossil animals and plants found on different continents. Unless such trans-oceanic land bridges had existed in the past these widely acknowledged similarities in former organic life were inexplicable in terms of Darwinian evolution. Genetic isolation should hence give rise to morphological divergences in the faunas of the different continents.

Suess gave the name *Gondwanaland* to a former continent embracing central and southern Africa, Madagascar, and peninsula India, after the late Palaeozoic *Gondwana* fauna common to its components. The term *Gondwanaland* has in its general usage been subsequently extended to embrace Australia, South America, and Antarctica as well. *Gondwana* (originally the name of a region in eastern India) is actually the more correct term, since it means land.

Suess also proposed the term *eustatic* for the world-wide rises and falls of sea level that could be inferred from the stratigraphic record of successive marine transgressions and regressions over the continents. He attributed the regressions to subsidence of the ocean basins by sediment from the continents. Water would therefore either be drained off the continents as the oceans deepened, or displaced onto them as a consequence of ocean-floor sedimentation.

While both Dana and Suess accepted the cooling contracting earth model, a fundamental difference was apparent at the end of the nineteenth century between European and North American interpretations. Whereas Europeans such as Suess and Emile Haug believed that extensive sectors of ocean were underlain by subsided continent, isostatic theory had more influence among the Americans, who were especially impressed by the fact that gravity surveys appeared to demonstrate that the oceans were underlain by denser crust than the continents. T. C. Chamberlin attempted to take the new geophysical data into account in his own theory, which combined elements of both contraction and isostatic theory; the continents and oceans were seen as permanent. Besides this evident contradiction in belief, it came to be increasingly agreed, from zones of overthrusting newly-recognized in the Alps, that the extent of lateral compression within mountain belts had been greatly underestimated, and certainly not predicted from contraction theory. Despite marked differences in opinions on these and other geotectonic matters, there was consensus on at least one issue; virtually all denied the possibility of significant lateral movement of continental masses through the oceans. In particular, any suggestion that present continents had drifted apart from a primordial landmass would have seemed decidedly heretical to the vast majority of geologists at the turn of this century.

Early notions of migrating continents

Prior to the mid-nineteenth century a number of people had noted the congruity of the coastlines of the South Atlantic, or had speculated that the Altantic Ocean had formed either by depression of a mysterious former continent 'Atlantis' or by excavation of an enormous valley. Snider was, however, the first to put forward the suggestion that the continents bordering the present Atlantic had formerly been adjacent and had then drifted apart.[4]

In 1858 Antonio Snider-Pellegrini published a book which in intellectual style and outlook bears more resemblance to a late seventeenth century speculative cosmogeny than a scientific treatise of the mid-nineteenth century.[5] Snider outlined the course of events between the Creation and the Deluge in terms of 'days' (epochs) consistent with the Genesis account. During the first day the freezing of a crust or a hot liquid interior caused extreme pressures to build up until the violent simultaneous explosion of numerous volcanoes blew the moon out of the earth. Four more epochs followed, each terminating with a cataclysm, until on the fifth day all the lands of the earth were concentrated in one large, unstable mass along which ran a giant fissure oriented approximately north–south. The Deluge occurred on the sixth day, as volcanic gases poured through the fissure, forcing the New and Old Word continents apart and causing a sudden contraction of the earth. Thus oceanic waters were forced over the continents and the Atlantic was born.

It was not to be expected that Snider's fantastic notions, put forward with hardly any supporting evidence, would be taken seriously by the geological community, especially as they relied heavily on the kind of catastrophist thinking that had been refuted so effectively by Lyell. Rupke[6] has indeed argued that the general reluctance to accept more respectable continental drift hypotheses might relate in part to its association with catastrophism.

Although his ideas did not receive general acceptance, the work of the Revd Osmond Fisher in England deserves mention, especially as his *Physics of the earth's crust*[7] is the first comprehensive geophysical treatise to be written. Fisher, like Dutton in America, who was the man most responsible for establishing the modern theory of isostatic compensation, developed serious doubts that contraction from cooling would be sufficient to cause the large amount of crustal shortening indicated by such features as the great folds and thrust faults that were beginning to be recognized in the Alps. As we have seen in the last chapter, he also challenged Kelvin's estimates of the age of the earth based on the cooling concept and postulated a relatively fluid interior with convection currents rising beneath the oceans – especially beneath the mid-Atlantic ridge – and falling beneath the continents. We may note here that this appears to be the first suggestion of a notion that is now universally accepted as an integral part of plate tectonics.

Fisher's claim that the continents and oceans were permanent derived from an appreciation of the principle of isostasy, but in his view they were only relatively permanent because he argued that as a result of convection the oceans must expand by the addition of volcanic rocks in a median position, and the continents must contract to form fold mountains at their margins. Furthermore, following G. H. (later Sir George) Darwin's proposal in 1879 that the moon had been expelled from the earth at an early stage in the latter's history, leaving behind the gigantic scar of the Pacific, Fisher indicated that one likely consequence would have been the lateral movement and fragmentation of the cooled granitic crust. Preoccupation with the alleged loss of the moon also figures in the later speculative hypotheses involving the opening of the Atlantic, put forward by Pickering[8] and Baker.[9]

In contrast to the by now firmly established Anglo-American tradition in geophysics, with its emphasis on the solid earth and the permanence of continents and ocean basins, the rather different German approach fully integrated meteorology and climatology into geophysics. While Wettstein's views were perhaps rather fanciful and his geotectonic interpretations excessively mobilistic, he at least made some attempt to draw support for his ideas from apparent palaeoclimatic and biogeographic anomalies in the stratigraphic record. Loeffelholz von Colberg[10] brought together much information on the 'southern continent' (Suess's Gondwanaland), past ice ages and other variations in the climatic and biological sequence in different parts of the earth, as evidence of a kind of crustal rotation (polar wandering). Kreichgauer[11] developed similar ideas more systematically and presented a series of palaeoequators and polar migration paths without differential continental displacement. The Austrian geologist Ampferer was much influenced by the discovery of radioactivity, with its significant implications for heat generation within the earth, and put forward a theory whereby convection currents in what we would now call the mantle caused contortions in the overlying crust.[12]

Thus at the turn of the century a mobilistic view of the earth, with segments of crust floating on a liquid interior and permitting polar wandering, was a familiar concept at least in Germany at a time when other physical studies of the solid earth were denying the possibility of polar wandering because of the apparent rigidity of the earth. The importance of this as a background to Wegener's own researches can hardly be over-estimated.

Before turning to Wegener there is one other important figure who must be mentioned, the American geographer and Pleistocene geologist F. B. Taylor, who in 1910 published a lengthy paper giving the first logically worked out and coherent hypothesis involving what we would now call continental drift.[13] It anticipated Wegener's writings on the subject by two years.[14]

The starting point of Taylor's hypothesis is not, as one might expect, the supposed fit of the continents bordering the Atlantic but the pattern of Tertiary mountain belts in Eurasia. On the eastern and southern borders of Asia and continuing into the Mediterranean region are a series of arcuate zones of mountains, usually convex towards the ocean, which exhibit signs of lateral compression in the form of folded or overthrust strata. Taylor found the conventional contraction hypothesis inadequate to explain satisfactorily the distribution and youth of the Tertiary mountain ranges. He envisaged a mighty creeping movement of the earth's crust from the north towards the periphery of Asia. The Indian Peninsula, an ancient shield area, acted as an obstructive block causing the huge pile-up of the Himalayas and Pamir plateau directly to the north while further east the fold ranges were able to swing round more freely, into Malaysia and Indonesia.

The notion of crustal creep from high to low latitudes in the northern hemisphere was supported in Taylor's paper by reference to Greenland, which was envisaged as the remnant of an old massif from which Canada and northern Europe had broken away along rifts. Suess and others has been struck by the close resemblance of Palaeozoic rocks and structures on the two sides of the North Atlantic but had attributed this to the collapse of Atlantis rather than the drifting apart of continental blocks. Certain points of detail in Taylor's interpretation are of considerable interest in the light of present knowledge. Thus the eastward curvature of the Scotia Arc, between Patagonia and the West Antarctic Peninsula, was seen as an indication that westward drift of the continental masses had lagged behind in this region.

Taylor paid little attention to the mechanisms of continental movement in his 1910 paper, but in subsequent papers suggested the operation of tidal forces when the moon was captured, rather than lost, by the earth during the Cretaceous. Unfortunately Taylor did not back up his interesting ideas with much evidence and his 'mechanism' involving the late capture of the moon must have seemed fantastic to his geological contemporaries. Consequently, his work had little impact, but it nevertheless ranks as the most important early contribution on continental drift apart from that of Wegener.

The continental drift of hypothesis of Alfred Wegener

To most people the notion of continental drift is irrevocably associated with the name of Alfred Wegener (1880–1930), because he was the first to put forward substantial evidence for a coherent and logically argued hypothesis that took account of a wide variety of natural phenomena.

Wegener was an outsider to the geological profession. Born in Berlin in 1880, the son of an evangelical minister, he studied at the universities of Heidelberg, Innsbruck, and Berlin and took his doctorate in astronomy. From his early days as a student he had cherished an ambition to explore in Greenland, and he had also become fascinated by the comparatively new science of meteorology. In preparation for expeditions to the Arctic he undertook a programme of arduous exercise. He also mastered the use of kites and balloons for making weather observations and was so successful as a balloonist that in 1906, with his brother Kurt, he established a world record with an uninterrupted flight of 52 hours.

Wegener was rewarded in his assiduous preparations by being chosen as a meteorologist to a Danish expedition to north-east Greenland. On returning to Germany he accepted a junior teaching position in meteorology at the University of Marburg and within a few years had written a textbook on the thermodynamics of the atmosphere. A second expedition to Greenland, with the Danish expeditioner J. P. Koch, took place in 1912; it is notable for the longest crossing by foot of the ice cap ever undertaken.

In 1913 Wegener married Else, the daughter of the meteorologist W. P. Köppen. After the First World War he succeeded his father-in-law as director of the Meteorological Research Department of the Marine Observatory at Hamburg. In 1926 he was finally offered and accepted a chair of meteorology and geophysics at the University of Graz in Austria; it had taken a long time to obtain adequate recognition for his wide-ranging contributions to science. Wegener died while leading a third expedition to Greenland in 1930, probably as a result of a heart attack. His laudatory obituaries concentrated on his great achievements as an Arctic explorer and pioneer meteorologist, whereas today of course he is best known as the most notable proponent of continental drift, though by his own admission this was for much of his professional lifetime little more than a peripheral interest.

There is no clear documentation as to how Wegener first conceived his hypothesis.[15] One unauthenticated story has it that he was inspired during his first trip to Greenland while watching the calving of glacier ice (the process by which icebergs are produced). By his own written account, however, the basic idea came to him in 1910, when he was struck by the remarkable congruence of the coastlines on either side of the Atlantic. He nevertheless initially treated the idea as improbable, but the next year he discoverd a report of palaeontological evidence for a former land bridge between Brazil and Africa. A search for more such evidence produced in his mind such weighty corroboration that he was led to develop the hypothesis which received its first public airing at a lecture in Frankfurt am Main in January 1912. There followed later that year two short publications entitled 'Die Entstehung der Kontinente'. An enlarged version first appeared in book form in 1915, as *Die Entstehung der Kontinente und Ozeane*. Revised editions appeared successively in 1920, 1922, and 1929. The fifth and sixth editions were published posthumously under the editorship of his brother Kurt, who added a memorial and further references. The third edition attracted much more attention than its predecessors and was translated in 1924 into English, French, Spanish, and Russian, the English version appearing under the title *The origin of continents and oceans*. In this edition Wegener's expression 'Die Verschiebung der Kontinente' was quite accurately translated as 'continental displacement'. Nevertheless the subsequently proposed term *continental drift* quickly became adopted in the English-speaking world.

The fourth edition represents the latest and most fully developed views that Wegener held, including some response to his early critics, and it is this edition which is most widely read today because it was reprinted in a new translation in English in 1966.[16] In this fourth edition both the character of the principal arguments and the way they are marshalled remain remarkably similar to those in his two papers from 17 years earlier, although they are greatly elaborated and there is much new evidence adduced, most notably from palaeoclimatology.

Interestingly enough, rather than outlining the 'jigsaw fit' and palaeontological clues which triggered the basic idea, he plunged directly into geophysical arguments, indicating what in his view were inadequacies or contradictions in orthodox theory. Only then did he cite geological evidence supporting the hypothesis that formerly united continents have rifted and moved apart.

Wegener pointed out that the cooling, contracting-earth model was vulnerable to attack on a number of grounds. In the first place, the comparatively recently discovered enormous overthrusted rock sheets or 'nappes' in the Alps led to estimates of Tertiary contraction which seemed excessive. Further, it was not clear from the contraction hypothesis why the shrinkage 'wrinkles' represented by fold mountains were not distributed uniformly rather than being confined to narrow zones. Further, some basic assumptions about the earth's supposed cooling, notably by Kelvin, had been undermined by the recent discovery of widespread radioactivity in rocks, leading to the production of considerable amounts of heat acting in opposition to thermal loss into space by radiation.[17]

Gravity data indicated that the ocean floor was underlain by denser rocks than on the continents, and the concept of isostasy rendered impossible the subsidence of vast continental areas into oceanic deeps, as some geologists still maintained.

Wegener postulated that, commencing in the Mesozoic and continuing up to the present, a huge supercontinent, 'Pangaea' (meaning *all land*), had rifted and the fragmented components had moved apart. South America and Africa began to separate in the Cretaceous, as did North America and Europe, but North America and Europe had retained contact in the north as late as the Quaternary. During the westward drift of the Americas, the western Cordilleran ranges had been produced by compression at or near the leading edges, but the Antilles and Scotia Arc had lagged behind in the Atlantic. The Indian Ocean had begun to open up in the Jurassic but the principal movements took place in the Cretaceous and Tertiary. A large area of land to the north of India had crumpled up in the path of India during its northward movement to form the Himalayas. Australia–New Guinea had severed its connection with Antarctica in the Eocene and moved northwards, driving into the Indonesian archipelago in the late Tertiary.

The principal supporting arguments and evidence are as follows.

1. Statistical analysis of the earth's topography reveals two predominant levels, corresponding to the surface of the continents and the abyssal ocean floor. Such a distribution would be expected in a crust made up of two layers, the upper one consisting of lighter rock such as granite, and the substratum consisting of basalt, gabbro, or peridotite, which would also form the ocean floor. It is not consistent with a model of the crust in which variations of elevation are the result of random uplift and subsidence, as widely postulated. In that case one would expect a Guassian, or bell-shaped, distribution of elevations around a single median level.

2. Well-established isostasy theory assumes that the substratum underlying the earth's crust acts as a fluid, albeit a highly viscous one. If the continental masses can move vertically through this substratum there is no reason why they should not also be able to move horizontally, provided only that there are forces of sufficient power to do this. That such forces do in fact exist is proved by horizontal compression of strata in mountain ranges such as the Alps, Himalayas, and Andes.

Wegener pointed out what he thought was a widespread confusion about the physical properties of the earth and drew an analogy with pitch, which shatters under a hammer blow like a brittle solid, but will in the course of time flow plastically under its own weight. In a similar way, the earth reacts as an elastic solid when acted on by short-period forces such as earthquake waves, but over the much longer periods signified by geological time it must behave more like a Newtonian fluid. For instance, the oblateness of the spheroid corresponds exactly to the period of rotation.

3. Wegener thought he could prove significant horizontal movement of Greenland away from Europe by means of geodetic observations using radio time transmissions. (We may note here that later, more refined, measurements discredited these data. The geodetic argument did not figure much in the subsequent controversy and no further reference will be made to it.)

4. The most cogent geological arguments concern the apparent similarities of rocks on the two sides of the Atlantic, suggesting a former continuity. The best kind of such evidence was the matching of orogenic fold belts. Wegener expresses the argument graphically in the following words.

It is just as if we were to refit the torn pieces of a newspaper by matching their edges and then check whether the lines of print run smoothly across. If they do, there is nothing left but to conclude that the pieces were in fact joined this way. If only one line was available for the test, we would still have found a high probability for the accuracy of fit, but if we have *n* lines, this probability is raised to the *n*th power.

Wegener went so far as to argue, by matching the terminal moraines of the European and North American ice sheets, that the two continents were not ruptured until as late as the Pleistocene.

5. The argument from palaeontology that is held to be one of the strongest is as follows. There are many late Palaeozoic and Mesozoic fossils in common to the southern continents which are now isolated by ocean. Basic biological principles demand a former free land communication to account for such a distribution, and this was traditionally explained by postulating transoceanic land bridges which sunk into the ocean floor some time after the Mesozoic. This explanation Wegener held to be untenable on geophysical grounds because it violated the principle of isostasy. The land bridge would be composed of granitic crust too light to sink into the denser rock of the sea floor. The only reasonable alternative explanation was that the continents had once been joined and had since drifted apart.

The point is made rather testily by Wegener in the following words.

A large proportion of today's biologists believe that it is immaterial whether one assumes sunken continental bridges or drift of continents – a perfectly preposterous attitude. Without any blind acceptance of unfamiliar ideas, it is possible for biologists to realize for themselves that the earth's crust must be made of less dense material than the core, and that, as a result, if the ocean floors were sunken continents and thus had the same thickness of lighter crustal material as the continents, then gravity measurements over the oceans would have to indicate the deficit in attractive force of a rock layer 4–5 km thick. Furthermore from the fact that this is not the case, but that just about the ordinary values of gravitational attraction obtain over ocean areas, biologists must be able to form the conclusion that the assumption of sunken continents should be restricted to continental shelf regions and coastal waters generally, but excluded when considering the large ocean basins.

Wegener attached most significance to the distribution of a small Permian reptile, *Mesosaurus*, known only from South Africa and southern Brazil, and the so-called *Glossopteris* flora, a Permo-Carboniferous flora found only in the southern continents and India, and long considered *prima facie* evidence for the entity called Gondwanaland by Suess. He also noted that the distribution of a number of living, non-marine organisms on the two sides of the Atlantic, notably certain earthworms, fresh-water fish, mussels, and land snails were difficult to account for if the Old and New Worlds had not had a land link at some time in the not-too-distant past.

6. The palaeoclimatic argument was based on the occurrence of distinctive rock types whose distribution was highly anomalous if the continents had remained in the same relative positions and if the poles had stayed fixed. Thus *tillites*, as found to occur widely in the southern continents, are pebbly mudstones formed from the melting of ice sheets, substantial *evaporite deposits* including rock salt and gypsum signify deposition in conditions of great warmth and aridity such as characterize large parts of the trade wind belt today, and *coals* with certain characteristic plant types signify humid equatorial conditions. The distribution of such deposits in the late Palaeozoic (Carboniferous and Permian) can only be satisfactorily accounted for by restoring the continents to the supercontinent Pangaea, and by invoking polar wandering.

Wegener was decidedly tentative on the question of the driving forces, as the following quotation indicates.

The Newton of drift theory has not yet appeared . . . It is probable that a complete solution of the problem of driving forces will still be a long time coming, for it means the unravelling of a whole tangle of interdependent phenomena, where it is often hard to distinguish what is cause and what is effect.

Nevertheless he could not resist a little speculation, probably assuming that even an inadequate and speculative working model would be better than none in establishing the plausibility of his hypothesis. Two possible components were recognized. A so-called *Pohlflucht* force (i.e. flight from the poles) was invoked to account for movement of continents towards the equator. This is a differential gravitational force, sometimes called the *Eötvös* force, resulting from the fact that the earth is an oblate spheroid, being relatively flattened at the poles. Another force must obviously be invoked for the westward drift of the American continents. Wegener favoured some kind of tidal force and argued that retardation of the earth's rotation by tidal friction must chiefly affect the outermost layers, which therefore should slide as a whole or as detached continental fragments over the interior.

Criticisms of Wegener's hypothesis

It appears that the initial reaction of the scientific community to Wegener's hypothesis was not uniformly hostile, but at best it was mixed. At the first lecture in Frankfurt some geologists were provoked to indignation; at Marburg, however, a few days later, the audience seems to have been more sympathetic. Following the early publications several prominent German geologists announced their opposition to the *Verschiebungstheorie*. A number of geophysicists, on the other hand, expressed approval of the concept. Indeed, in 1922, according to his widow's biographical account, Wegener was able to say that he knew of no geophysicist who opposed his hypothesis. This seems remarkable in view of the fact that subsequently geophysicists were to be among the most vehement critics.

The response in England to the appearance of the third edition of Wegener's book can be judged from a critical review by Philip Lake,[18] and also by his comments at a meeting of the Royal Geographical Society in 1923. (The previous year there had been a lively debate on the subject at the annual meeting of the British Association.)

In examining the ideas so novel as those of Wegener it is not easy to avoid bias. A moving continent is as strange to us as moving earth was to our ancestors, and we may be as prejudiced as they were. On the other hand, if continents have moved many former difficulties disappear, and we may be tempted to forget the difficulties of the theory itself and the imperfection of the evidence... Wegener himself does not assist his reader to form an impartial judgement. Whatever his own attitude may have been originally, in his book he is not seeking truth; he is advocating a cause, and is blind to every fact and argument that tells against it. Nevertheless, he is a skilful advocate and presents an interesting case.

Having made this concession Lake goes on to attempt to demolish the case. For instance, he attacks Wegener's argument that the continents and ocean floor represent two distinct layers, on the basis of the bimodal topographic distribution. The tectonic deformation of a single surface would not, according to Wegener, produce two maxima unless some additional unspecified physical causes intervened. Since this is not the case the frequency should be controlled by Gauss's Law of Errors. Lake argues that Wegener's reasoning leads us

to the remarkable principle: if we do not know the law, the law must be the law of errors... And so the actual heights [of the earth's surface] must be influenced by the extent of our knowledge. But Wegener's world is not an ordinary one. In his diagram he draws a broken line, and this, he says, follows approximately the course of the law of errors, according to which the frequencies would be regulated if one level only had been involved in the subsequent movements. The frequencies in the diagram are expressed as percentages, and in the ordinary world would add up to 100. Wegener's line gives a total of about 200... A mere inspection of the diagram is enough to show that the two curves cannot both be correct, for the areas contained between them and the vertical axis should be equal.

Lake cited an analysis which apparently showed that undulations of a single crustal level would of necessity produce a bimodal hypsometric curve. He went on to criticize other aspects of Wegener's geophysics, and was particularly scathing about his attempted fit of the continents. It was apparent that Wegener's reconstruction of Pangaea involved considerable distortion.

It is easy to fit the pieces of a puzzle together if you distort their shape, but when you have done so, your success is no proof that you have placed them in their original positions. It is not even a proof that the pieces belong to the same puzzle, or that all of the pieces are present.

Furthermore, Lake doubted the reliability of Wegener's matching of geological features from continent to continent and expressed extreme scepticism about both the proposed post-Pleistocene opening of the North Atlantic and the geodetic measurements allegedly indicating rapid westward movement of Greenland away from Europe.

Among the speakers who contributed to the discussion following Lake's lecture to the Royal Geographical Society G. W. Lamplugh was sceptical but sympathetic.

It may seem surprising that we should seriously discuss a theory which is so vulnerable in almost every statement as this of Wegener's... But the underlying idea that the continents may not be fixed has in its favour certain facts which give every geologist a predilection towards it in spite of Wegener's failure to prove it.

The geophysicist R. D. Oldham was anxious to establish that Wegener's hypothesis had been anticipated by an Englishman. He referred to his early days as a scientist.

But also I remember very well that in those days it was unsafe for anyone to advocate an idea of that sort. The physicists, who before that had forced on us the notion of a fiery globe with a molten interior and thin crust on it, had gone round and insisted on a solid heated sphere, and they would allow us to appeal to nothing, as the cause of various structures and changes that we knew in geology, but the slow cooling and contraction of this solid globe, and any notion of the shifting of continents was incompatible with that theory. Those ideas held the ground so strongly that it was more than any man who valued his reputation for scientific sanity ought to venture on to advocate anything like this theory that Wegener has nowadays been able to put forward...

But there was one man, even then, who did quite formally propose and maintain something ... almost identical with the Wegener hypothesis. That was Osmond Fisher, a man who in his time was a scientific Ishmaelite.

The consensus of the meeting appeared to be that, while there were some difficulties about the contracting-earth model and the doctrine of oceanic permanence, Wegener's hypothesis posed no serious threat to conventional beliefs. One of the strongest critics in the discussion was Dr (later Sir Harold) Jeffreys (1891–1989) who became renowned as a great pioneer of mathematical geophysics and one of the most influential and vehement critics of the continental-drift hypothesis.

Jeffreys went to Cambridge University to study mathematics and spent the rest of his life there, being elected to the Plumian Professorship of Astronomy in 1946. He made many distinguished contributions both to mathematics and to geophysics. While he had contributed much to the understanding of tidal friction, the earth's thermal history, and stresses in the crust, his work in seismology has proved the most fundamental, notably his deductions of the times of travel of earthquake waves. His knowledge of how great earthquake shocks caused the earth to 'ring like a bell' undoubtedly influenced his view that the earth must possess considerable strength, reacting as an elastic solid to such sudden releases of accumulated stress.

Jeffreys[19] expanded on his criticisms of drift in his great geophysical treatise which appeared in 1926.

A further impossible hypothesis has often been associated with hypotheses of continental drift and with other geological hypotheses based on the conception of the earth as devoid of strength. That is, that a small force cannot only produce indefinitely great movement, given a long enough time, but that it can overcome a force many times greater acting in the opposite direction for the same time. In Wegener's theory, for instance ... the assumption that the earth can be deformed indefinitely by small forces, provided only that they act long enough, is therefore a very dangerous one, and liable to lead to serious error.[20]

Jeffreys argued that if the sima is the weaker layer and will allow continents to plough through it like ships sailing before the wind, then it will not crumple their prows – even less will it crumple their prows and let them keep sailing. Yet Wegener required precisely this to account for his cordilleran mountain belts.

However there was no evidence that the sima was even the weaker medium. The continental crust is strong enough to support Mount Everest, and the oceanic crust strong enough to hold down deep trenches. The Eötvös ('Pohlflucht') and tidal forces postulated by Wegener can be shown by simple calculations to be only about one millionth as powerful as that required to move continents. Jeffreys' own interpretation of our planet involved an updated version of the contracting-earth model.

Jeffrey's criticisms mark the start of perhaps the most formidable opposition that Wegener faced, what might be called the ultrasolid-earth school of geophysicists in Great Britain and the United States, who believed they could demonstrate definitely that the earth possessed too great a strength to allow continents to migrate across its surface.

The second edition of Jeffreys' book, which appeared in 1927, adds the statement that the African and South American coastlines will not fit. This is undoubtedly true, but the fit at the outer margin of the continental shelf, which approximates more closely to the true continental edge, was later shown to be excellent. Wegener's continental outlines in fact relate to this geologically more significant boundary. Thus the Falkland Plateau of the South Atlantic is treated by Wegener as part of the continent.

The symposium held in New York in 1926, sponsored by the American Association of Petroleum Geologists, the proceedings of which were published two years later,[21] ranks highly in the earlier debate about continental drift. It was the first international meeting devoted entirely to the subject, and both Wegener and Taylor attended. Judging from the published proceedings, the only person with any real conviction of the truth of continental drift, apart obviously from the two above-mentioned persons, was the organizer and publication editor, the Dutch oil geologist van Waterschoot van der Gracht.

C. R. Longwell, of Yale University, was at least sympathetic, but unconvinced by the evidence put forward by Wegener.

Perhaps the very completeness of the iconoclasm, this rebellion against the established order, has served to gain for the new hypothesis a place in the sun. Its daring and spectacular character appeals to the imagination both of the layman and of the scientist. But an idea that concerns so closely the most fundamental principles of our science must have a sounder basis than imaginative appeal.

Other contributions were more hostile. Apart from the editor's, the longest article was written by Charles Schuchert, another Yale professor and one of the leading North American experts on reconstruction of ancient palaeogeographies. He was one of the few to address himself to the ancient faunal and floral similarities between continents. Far from denying them, Schuchert strongly supported them, but argued that they were in fact perfectly explicable by means of invoking a few land bridges, such as across the Arctic, the South Atlantic, and the Indian Ocean, which subsided after the Mesozoic. Furthermore, the different faunas were similar but not identical, which one would expect if they had formerly been contiguous. He was particularly caustic about Wegener's interpretation of the *Glossopteris* flora.

Any palaeontologist who reads carefully pages 98 to 106 of Wegener's book, dealing with the distribution of the Coal Measures and *Glossopteris* floras of 'Permo-Carboniferous Time', will see not only the nimbleness and versatility of his mind, but as well how very easy it is for him to make all from the generalizations of others, and compares unlike things, regarding the correlation of formations by geologists as dealing with 'relatively trifling differences of time'. In these pages he is explaining his views of the climate of Permo-Carboniferous time and, in doing so, shoves the south pole to a place off the southeast coast of Africa, arranging the equator accordingly. Finally, to make it easy for all of us to get his views, he pictures them on a single diagram entitled 'Evidences of Climate in the Permo-Carboniferous'. This single diagram undertakes to represent events that took place during a lapse of something like 50 million years, makes the flora of the tropical Coal Measures fit the 'polar' *Glossopteris* flora of the much younger Permian, and in order that the latter may be truly polar assumes that it was treeless, says the Antarctic then was adjacent to south eastern Africa with the south pole at the edge of it, and on this basis arranges the climatic belts around it!

Schuchert was perhaps the first to make an often-repeated criticism. Why had Pangaea endured for so much of geological history as a coherent supercontinent and then suddenly started to split up a mere 100 million years ago? If the Tertiary mountain ranges (Alps, Himalayas, Rockies, Andes, etc.) owed their origin to drift, what had produced the many much older mountain ranges, such as the Caledonian and Appalachian fold belts?

Schuchert also challenged the accuracy of Wegener's continental fits, and obtained very different results from his own experiments with plasticine on a model globe. These convinced him that no significant matching of continental margins was possible without considerable distortion. In particular, any scheme of fitting the Americas to Europe and Africa created a large gap at the Bering Strait, whereas the consensus of palaeontologists was strongly in favour of a long-operating if intermittent land bridge there. Scepticism was also expressed about the geological similarities between continents, claimed by Wegener, and about the supposed palaeoclimatic evidence for polar wandering.

A leading American structural geologist, Bailey Willis, questioned Wegener's geophysics and made a point similar to one put forward by Jeffreys. How could the sima be the more yielding crustal layer and yet resist the sialic rafts of the continents until cordilleran mountain ranges are produced by backthrust? The American geophysicist William Bowie also backed up another of Jeffrey's criticisms. If the sima is a weak layer how could the ocean floors maintain ridges and trenches instead of smoothing out like a mud flat? Polar wandering was also dismissed as geophysically impossible.

Some contributors could not resist questioning Wegener's respectability as a scientist. Consider for instance the comment of the American palaeontologist E. W. Berry.

[Wegener's method] in my opinion, is not scientific, but takes the familiar course of an initial idea, a selective search through the literature for corroborative evidence, ignoring most of the facts that are opposed to the idea, and ending in a state of auto-intoxication in which the subjective idea comes to be considered as an objective fact.

Quite the most outspoken of Wegener's critics was the American geologist R. T. Chamberlin, son of Kelvin's adversary, T. C. Chamberlin.

Wegener's hypothesis in general is of the foot-loose type, in that it takes considerable liberty with our globe, and is less bound by restrictions or tied down by awkward, ugly facts than most of its rival theories. His appeal seems to lie in the fact that it plays a game in which there are few restrictive rules and no sharply drawn code of conduct.

Chamberlin listed no fewer than 18 points which he considered destructive of the drift hypothesis. Many of these seem downright foolish today, but as an Ice Age expert he made the telling point that Wegener's matching of moraines to prove continental linkage in the Quaternary was quite ludicrous. Another glacial expert, A. P. Coleman, had in 1925 written an article criticizing the late Palaeozoic ice sheet inferred for Gondwanaland, making the point that much of the area in question would have been far from the ocean on Wegener's reconstruction, and hence out of reach of moisture-laden winds. Such a region would be more like Central Asia, which never receives heavy snowfalls despite having the coldest temperatures recorded on earth.[22]

All of the early debate concerning the pros and cons of the continental drift hypothesis was conducted in almost complete ignorance of the geology of the ocean floor. It was therefore not possible to dismiss out of hand the views of a distinguished geologist, J. W. Gregory, who argued, in a presidential address to the Geological Society of London,[23] for an elaborate system of sunken continents beneath the ocean which rose to the surface periodically, to provide trans-oceanic land bridges for inter-continental migrations. To believe this he felt he had to dispose of isostasy.

If the ocean surface does not conform to a regular ellipsoid or spheroid, if it sags down in mid-ocean owing to the lateral attraction of the water toward the land, or sinks with variations in the specific gravity of the water ... then the slight differences in the attraction of the ocean floor may be due to the depth being overestimated and not to the higher density of the floor.

He concludes: 'If isostasy is so stated that it is inconsistent with the subsidence of the ocean floor, so much the worse for that kind of isostasy'.

Decisive recognition that remarks of this sort were nonsense would have to await the intensive exploration of the ocean floor which began after the Second World War and which totally transformed our knowledge of the earth.

Wegener's response to his critics

Else Wegener's biography of her husband gives us some fascinating documentation of Wegener's private thoughts in the early years concerning the critical reaction to his hypothesis. Evidently Köppen was the earliest sceptic, and had attempted to discourage his son-in-law from straying from the straight and narrow path of meteorology to speculate on large subjects in which he was unqualified. Wegener wrote the following letter in reply in January, 1911, that is, before the new ideas were in the public domain.

I believe that you consider my primordial continent to be a figment of my imagination, but it is only a question of interpretation of observations. I came to the idea on the grounds of the matching coast-lines but the proof must come from geological observations. These compel us to infer, for example, a land connection between South America and Africa. This can be explained in two ways: the sinking of a connecting continent or separation. Previously, people have considered only the former and have ignored the latter possibility. But the modern teaching of isostasy, and more generally our current geophysical ideas, oppose the sinking of a continent because it is lighter than the material on which it rests. Thus we are forced to consider the alternative interpretation. And if we now find many surprising simplifications and can begin at last to make real sense of an entire mass of geological data, why should we delay in throwing the old concept overboard. Is this revolutionary? *I don't believe that the old ideas have more than a decade to live.*[24] At present the notion of isostasy is not yet thoroughly worked out; when it is, the contradictions involved in the old ideas will be fully exposed.[25]

To Wegener's clear, logical mind it was evidently all rather obvious and straightforward. It is intriguing to learn that he eschewed the notion that what he was about to propose to the scientific public was especially revolutionary. Quite evidently he seriously underestimated the strength of the critical reaction that was to follow, because the 'old ideas' persisted much longer than a further decade.

Wegener's response to German critics in the years immediately following the First World War is indicated in another letter to his father-in-law.

Professor P.'s letter is typical! He will not allow himself to be taught. Those people who insist in treating only with the facts and want to having nothing to do with hypotheses, themselves are utilising a false hypothesis without appreciating it!... there is nothing in his letter about the struggle to get to the bottom of things, but only about the pleasure of exposing the limitations of other men. Such men will have nothing to do with a reorientation of ideas. If they had learned the displacement theory at school they would uphold it uncritically with the same lack of understanding as they now do with the sinking of the continents into the oceans...[26]

This letter betrays Wegener's exasperation at what he evidently regarded as the incorrigible conservatism and resistance to new ideas of many well established geologists. This exasperation must have increased through the course of the 1920s as knowledge of his work spread through the scientific world, and the groundswell of opposition increased. Nevertheless he made no particularly forceful attempt to reply to his many critics in the fourth edition of his book, perhaps because in the late 1920s he was heavily preoccupied with administrative matters, in particular the organization of the fateful 1930 expedition to Greenland.

His main response was to attack the partiality of his critics, pointing out that geology is an integrative, synthetic discipline, in which conceptual models must take account of evidence from a wide range of subjects. 'Scientists still do not appear to understand sufficiently that all earth sciences must contribute evidence towards unveiling the state of our planet in earlier times, and that the truth of the matter can only be reached by combining all this evidence.' Elsewhere in his book Wegener expresses a similar attitude towards narrow-minded, blinkered partiality by means of the following graphic simile.

We are like a judge confronted by a defendant who declines to answer, and we must determine the truth from the circumstantial evidence. All the proofs we can muster have the deceptive character of this type of evidence. How would we assess a judge who based his decision on part of the available data only?

Wegener's early supporters

It is ironic in view of later developments that, whereas certain leading geologists in the German-speaking world, such as Diener, Soergel, and Semper, expressed their strong opposition to the continental-drift hypothesis when it was first presented, the view of at least some geophysicists was sympathetic. Thus on 24 September 1912 Köppen wrote in a letter to his daughter: 'Yesterday, on the occasion of the International Geodetic Meeting, I discussed Alfred's continental hypothesis with the geodesists Helmert, Albrecht and Förster. I was delighted to learn how seriously they take it. In particular, old Albrecht considered it to be a very interesting idea that needed to be put to the test.'[27] This sympathetic interest, if not active support, of German-speaking geophysicists continued into the years following the First World War. Consider, for instance, the following extract from a letter written to Wegener by Milankovitch, who had become acquainted with Wegener and Köppen as a result of their common interest in palaeoclimates; his bias as a mathematical physicist is evident.

I am still totally under the influence of your brilliant lecture. It had such a powerful impression because it was so closely argued and confined itself to the most important facts. You had indeed told me that I am easily convinced. That is certainly true when such factual material comes under the embrace of a general idea. I am not in the least bit disturbed that not all geological details will fit neatly into your picture. The same applies to those who occupy themselves with the investigation of the mechanism behind complex natural phenomena. Those who on the contrary have spent their whole lives collecting and recording facts, are incapable of penetrating beyond these facts; their view is fixed on the surface. You should not be discouraged if you find it more difficult to persuade empiricists than students of the exact sciences, quite the contrary in fact.[28]

As regards those early supporters of some form of continental mobilism who were prepared to commit themselves to print, the number is unsurprisingly small, but significantly includes some of the outstanding geologists of the time. A number of Swiss, French, and Dutch structural geologists working respectively in the Alps, the Moroccan Atlas Mountains, and Indonesia, were impressed with the magnitude of crustal motion indicated by nappe structures and overfolds.

The greatest of these was the Swiss geologist Emile Argand (1879–1940), one of the founders and principal developers of the nappe theory of the Alps. Argand presented his views on the subject in 1922 at the 13th International Geological Congress in Brussels. The published work[29] is a lengthy synthesis of global tectonics, in which the evolution of the earth's crust is discussed in terms of '*mobilisme*' as opposed to '*fixisme*'. Argand favoured Wegener's concept of floating continents and of plasticity under long-term stress of rock materials. Fixism was not an adequate theory at all whereas Wegener's mobilism was a comprehensive theory supported by a large and diverse body of evidence to which Argand added many details from his studies of fold mountains. He also pointed out that Wegener had raised many important issues which his opponents had failed to explain away. One should also note that Suess's son Franz Edvard, Professor of Geology in Vienna, was a drift supporter whose mobilistic views on the formation of mountain belts show a remarkable anticipation of modern thought, for example the obduction of ophiolite belts.

Since the United States quickly developed into and long remained the centre of strongest opposition to continental drift, it must have taken the great American geologist R. A. Daly (1871–1957) more than a little courage to put forward mobilistic views in his book *Our mobile earth*.[30] While he fully accepted that drift had occurred he had serious reservations about the proposed mechanism, and suggested an alternative whereby the continents slid laterally under the influence of gravity owing to bulging of the polar and equatorial regions with a depression in between. This he called his 'downsliding' or 'landsliding' hypothesis.

The continents appear to have slid downhill, to have been pulled down under the earth's body by mere gravity; mountain structures appear to be the product of enormous slow *landslides*. Each chain has been folded at the foot of a crust-block of continental dimensions which was not quite level, but slightly tilted.

The tilting was thought to be due to the earth's contraction and changing speed of rotation, and to erosion of land, leading to increase of load in the marine basins. Distortions alone, however, would never lead to continental sliding unless there were a zone of weakness underlying the earth's strong outer crust. The notion of a weaker, deeper zone underlying a stronger, more rigid shallow zone is, incidentally, paralleled in the modern concepts of the asthenosphere and lithosphere.

According to Daly's early ideas continental sliding has taken place throughout geological history and has caused the creation of major mountain ranges. He thought that by late Palaeozoic times, after a long history of motion and growth, all the landmasses had coalesced into a single supercontinent that subsequently commenced breaking up in the Mesozoic. This anticipates modern views on the formation of Pangaea, a subject to which Wegener did not address himself.

In later years Daly back-tracked to a considerable extent from his extreme mobilistic views of the 1920s but persisted in his belief that gravitational sliding was the principal cause of mountain building.

In Great Britain the outstanding structural geologist E. B. Bailey was favourably disposed towards the drift hypothesis because it appeared to account so well for the similarities and crossing of the Caledonian and Hercynian fold belts on the two sides of the North Atlantic. Another British geologist, Arthur Holmes, played a far more significant role in promoting the drift hypothesis, which he strengthened considerably by proposing a far more plausible mechanism of continental movement than those put forward by Wegener.[31]

Besides his work on establishing an absolute time-scale, mentioned in Chapter 5, Holmes had also achieved a considerable reputation for his research in igneous petrology, and had in 1925 been appointed to the chair of geology at Durham University. The thermal effects of radio-activity played a prominent role in a lecture he gave to the Geological Society of Glasgow in 1928. An expanded version of the lecture, published three years later,[32] ranks as a fascinatingly prophetic work.

Holmes proposed a model of the earth with upper, intermediate, and lower layers composed respectively of granitic, dioritic, and peridotitic (plus eclogitic) rocks. The *crust* consisted of the upper and intermediate layers together with the higher, crystalline part of the lower layer, and the *substratum* of the lower, glassy or thermally 'fluid' part of the lower layer. The sub-sedimentary basement of the ocean floor was composed of the plutonic rock, chemically equivalent to basalt, known as gabbro, or amphibolite, a metamorphosed equivalent of gabbro.

It had become evident by the late 1920s that ordinary volcanic activity as observed was insufficient to discharge the amount of heat from radioactivity in the substratum that was estimated to rise to the earth's surface. The data made more sense if one invoked convection currents in the substratum which, if rising beneath continents, could cause continental drift. Holmes pointed out that at high temperatures the strength of materials rapidly diminishes, and consequently the substratum should be devoid of strength, contrary to what Jeffreys had argued. The critical gradient at which convective circulation should commence was estimated to be about 3 °C/km, and calculations suggested that the viscosity of the substratum is not too high for convection to occur.

Because of the earth's rotation, upward-moving currents would be deflected westwards and downward-moving currents eastwards. 'Monsoon-like' currents might be expected to occur owing to the distribution of continental blocks and ocean floors. The radioactivity of continental rocks is higher (because uranium and thorium are concentrated in granites and the temperature beneath them should be higher than under the ocean). Currents would therefore rise under the continents and spread in all directions towards the peripheral regions. Above the places where the currents rise and diverge a stretched region would arise in the continental crust and eventually fragments would be torn asunder, leaving a disruptive basin to subside as new ocean, into which much excess heat would be discharged.

The upper granitic layer of the crust would be thickened by differential flowage of its levels towards the obstructing ocean floor. This crustal thickening at the leading edge of the moving continents would cause mountain building. The mountain roots, being light and therefore unable to sink, would begin to fuse and give rise to the andesites and associated volcanic rocks such as those found so commonly in the so-called 'Fiery Ring' of the Pacific borderlands. This mechanism for mountain building associated with continental drift avoids the difficulty pointed out by critics of Wegener, with the 'sima' being weaker than the advancing 'sial' but nevertheless able to exert a backthrust. By Holmes' proposed mechanism, mountain building would inevitably occur provided that the horizontal component of rock flowage was greater from behind than in front.

Holmes went on to outline a number of other geological consequences, and indicated that his hypothesis accounted plausibly for such major phenomena as geosynclines and the East African rift valley.

South Africa was the home of one of the leading field geologists of his day, Alex du Toit (1878–1948). Apart from acquiring an impressive knowledge of his native country, he had been struck by the remarkably close resemblance of the Palaeozoic and Mesozoic geology of South Africa with eastern South America. Wegener saw du Toit's lengthy article on this subject, published in 1927, and referred to it enthusiastically in the fourth edition of his book. Du Toit became a leading disciple of Wegener and propounded drift theory indefatigably until his death. He is best known for his book *Our wandering continents*.[33]

Du Toit presented a wealth of new evidence favouring drift, chiefly concerning geological comparisons of the southern continents about which the geologists of Europe and North America were largely ignorant, and added some interesting new arguments. Of particular value is his attempt to eliminate some of Wegener's weaker points such as postponing final separation of North America from Europe until the late Pleistocene, on the basis of the matching of terminal moraines. A more accurate fit of the continents on the edge of the shelves was attempted, while the frequently expressed objection to Wegener, that his hypothesis failed to account for the formation of the pre-Tertiary mountain belts, is countered by arguing that the Caledonian and Hercynian mountain-building episodes of the Palaeozoic also involved continental drift of a sort.

Rather than accepting Wegener's single supercontinent Pangaea, du Toit preferred to argue for northern and southern supercontinents (Laurasia and Gondwanaland, respectively) which had been separated since the late Palaeozoic by the seaway Tethys (a term of Suess) which was not broken up until the Tertiary as Africa and India drove northwards towards Eurasia.

Gondwanaland was reconstructed with far more accuracy than Wegener attempted, and du Toit had a great deal to say about a host of other matters, some of it highly percipient. Thus his statement that, during Eocene times, the Iberian Peninsula was rotated in an anticlockwise direction, opening up the Bay of Biscay and causing compressional movements (folding and thrusting) in the Pyrenees, anticipates modern views, based on much more evidence, quite closely.

The community of biogeographers appeared to be split between proponents of land bridges (adhered to despite Wegener's cogent criticism) and drift. Among the latter, European phytogeographers were perhaps the most sympathtic. Perhaps the leading one to opt for the drift solution, notably with reference to the *Glossopteris* flora, was the British palaeobotanist A. C. Seward. In his 1924 Presidential Address to the Geological Society of London he stated the following.

Although it would be presumption on my part to attempt to discuss the possibility of finding assistance in our endeavour to recreate Gondwanaland in the much discussed hypothesis of Wegener, I must confess some sympathy with Dr. du Toit's acceptance of the view that the present South America, South Africa, India and Australia represent portions of the ancient continents finally torn apart...

In his 1929 Presidential Address to section K (botany) of the British Association he reasserted his sympathy for the drift solution, which he further elaborated on as his support grew stronger in his *Plant life through the ages*, published in 1931. He was especially impressed by the association of the *Glossopteris* flora with the Gond-wana ice deposits. Several of Seward's students who became leading botanists or palaeobotanists in later years were also staunch suppor-ters of drift, notably B. Sahni and R. Good.

Later criticisms of drift

It might be thought that the combined efforts of Holmes and du Toit, the one putting forward a plausible mechanism, the other introducing a wealth of new supporting geological evidence, might have at least begun to sway opinion towards Wegener by the time of the Second World War. In particular, the old orthodox model of a cooling, contracting earth had been dealt a severe blow by the discovery of radio-activity, and geologists could no longer plead ignorance of the southern continents, with their remarkably widespread tillite deposits suggesting an enormous amount of ice cover in the late Palaeozoic.

On the contrary, opinion seems to have hardened against the notion of drift. Jeffreys had responded critically to Holmes' convection hypothesis in a discussion of the Geography Section of the British Association in 1931.

I have examined Professor Holmes's theory of subcrustal currents to some extent, and have not found any test that appears decisive for or against. So far as I can see there is nothing inherently impossible in it, but the association of the conditions that would be required to make it work would be rather in the nature of a fluke... For the theory to succeed would require the currents to be in the same direction as regions of continental dimensions; that is, the instability developed must correspond to the lower mode of disturbance and no other.

Although it cannot be proved it appears likely that Jeffreys' great prestige as a geophysicist was such that a critical word from him would carry disproportionate weight.

The response to Wegener in Europe was hardly more sympathetic.

After some initial expressions of sympathy and interest the consensus turned against him. Without much doubt one of the most significant influences in this respect during the inter-war years was the great German tectonician Hans Stille, who had lectured and written for several decades on his theory of a contracting earth subject to a succession of brief global episodes of mountain building generated by geosynclinal compression, interrupted by longer episodes of tectonic quiescence. There was no room in Stille's scheme for significant lateral as opposed to vertical mobility, and his formidable authority carried great weight not just in his home country but in North America, where Walter Bucher, an equally influential tectonician, supported a comparable global scheme that was just as staunchly stabilist.

The American palaeontologist G. G. Simpson noted in 1943[34] the near unanimity of palaeontologists against Wegener's ideas. As regards his own special interest of fossil mammals he was quite categoric. 'The known past and present distribution of land mammals cannot be explained by the hypothesis of drifting continents ... the distribution of mammals definitely supports the hypothesis that the continents were essentially stable throughout the whole time involved in mammalian history.'

Simpson criticized the interpretations of both Wegener and du Toit, for instance concerning the isolation by drift of Madagascar. Simpson's forceful and cogent paper appears to have been very influential in America, the more so because he was able to dismiss the land-bridge alternative to drift, in favour of chance crossings by animals of an ocean, what he termed 'sweepstakes routes'.[35]

Although there were always some biogeographers who favoured drift, the quality of their evidence and argumentation was often not high and they were not taken seriously by most palaeontologists. In 1949 a symposium was held in New York, by the Society for the Study of Evolution, to be pubished three years later in the *Bulletin of the American Museum of Natural History*. The subject was the problem of land connections across the South Atlantic. Of the 17 participants only three offered support for drift.

The venerable Bailey Willis was even more outspoken than Simpson.

I confess that my reason refuses to consider 'continental drift' possible. This position is not assumed on impulse. It is one established by 20 years of study of the problem of former continental connections as presented by Wegener, Taylor, Schuchert, du Toit, and others with a definite purpose of giving due consideration to every hypothesis which may explain the proven facts. But when conclusive negative evidence regarding any hypothesis is available,

that hypothesis should, in my judgement, be placed in the discard, since
further discussion of it merely incumbers the literature and befogs the mind
of fellow students... Fellow scientists who are not geologists cannot be
expected to know that the geology upon which protagonists of the theory rest
assumptions is as antiquated as pre-Curie physics. Wegener and his successors
are disciples of Eduard Suess, the Master of European geologists. I knew him
well: a charming, genial German, who never travelled far, but assembled the
observations of others and from them constructed speculations regarding the
face of the Earth. His reading was prodigious, his memory marvellous, his
imagination grand; but he gravely lacked critical faculty. And when some
airy concept had grown in his mind, it became too firmly rooted ever to be
dislodged.

Such a concept was Gondwana Land, the continent supposed to have
extended from the East Indies westward to the Pacific, embracing India,
Africa and South America and occupying the sites of the Indian and South
Atlantic oceans. It had no actual existence...

Thus the theory of continental drift is a fairy tale, ein Märchen. It is a
fascinating fancy which has captured imaginations.[36]

Only a geologist assured of widespread support would have ventured
such extreme and patronizing views in print. As late as 1950 Gevers,
in his otherwise laudatory obituary of du Toit, felt constrained to half
dissociate himself from heresy.

Notwithstanding the zealous and valiant efforts of du Toit and others, there
has in recent years been a marked regression of opinion away from
continental drift... The greater the indignation to which the orthodox minds
are roused by revolutionary heresies, the greater the amount of unsympathe-
tic attention. This again stimulates the zeal and ardour of the heretics and
their disciples... Of late, however, the obstacles to smooth continental
drifting are being more strongly felt in many quarters previously
sympathetic.[37]

New research in geophysics and oceanography

After the Second World War the application of new or more refined
geophysical techniques began to transform our knowledge of the earth
in a radical way, and these were eventually to lead to a profound
revolution in thought from a stabilist to a mobilist view. The most
significant development was the greatly intensified study of rock
magnetism.

Knowledge of rock magnetism goes back to classical times but
nothing relevant to the subject under study was done till the work of
Brunhes early this century. He and his successors showed that: (1)
modern lavas are magnetized in the direction of the present field; (2)

Tertiary lavas are magnetized roughly in the direction of the present field or in the reverse direction; and (3) older lavas are often magnetized in directions making large angles with the present field.[38] It is remarkable that such fascinating and obviously significant phenomena should have been so widely ignored for so many years.

A major breakthrough in modern research came with the development of a highly sensitive magnetometer by the British physicist, P. M. S. Blackett, which made the study of a wider range of rock samples more practicable.

The magnetization of rocks is in effect a fossil permanent magnetism which can serve as a kind of fossil compass to determine the direction of the ancient or palaeomagnetic field. The research method is based on the assumption that the mean geomagnetic field is that of an axial dipole situated at the earth's centre. Oriented rock samples of certain types yield information on the field direction (giving information about the direction of the poles) and latitude, if the angle of declination D and the angle of inclination I are determined by a sensitive magnetometer. The angle of latitude L is determinable from the simple equation,

$$\tan I = 2 \tan L$$

Only a few types of rock can be used. Basaltic lavas are fairly iron-rich and acquire their magnetization from the geomagnetic field as they cool through the Curie points of their iron-titanium oxide minerals upon crystallization after they have erupted at the earth's surface.

Certain sedimentary rocks such as red sandstones contain enough iron oxides to acquire a measurable magnetization. The way in which such sediments become magnetized remains somewhat obscure. To some extent grains of iron oxide minerals may behave as minute magnets during sedimentation and align themselves in the field. Magnetization may also be acquired subsequent to deposition, however, during the process of chemical and physical alteration known as diagenesis.

The fundamental assumption concerning the axial nature of the field is justifiable on both theoretical and empirical grounds. It is consistent with the widely accepted theory of the geomagnetic field put forward by W. M. Elsasser and Sir Edward Bullard, which argues that the fluid, electrically-conducting outer core of the earth acts as a self-exciting dynamo. Although at the present time the magnetic axis deviates by several degress from the geographic (rotation) axis it is virtually coincident when averaged over periods of several thousand years. Furthermore, palaeomagnetic determinations on lava samples going back 10 million years or so into the Upper Tertiary indicates a position for the geographic North Pole that cannot be distinguished statistically from that of today.

By the middle 1950s some intriguing results were being obtained by research groups at Cambridge and Imperial College, London. The Cambridge group, notably S. K. Runcorn, K. M. Creer, and E. Irving, obtained samples of European rocks of widely different ages and were able to demonstrate a steady change with time, prior to the later Tertiary, of the position of the North Pole from a late Precambrian position near Hawaii. Comparable results for other continents were obtained by Blackett's group at Imperial College.

As the differences compared with the present were not random but systematic, with the angle of polar difference increasing with time, a migration either of the poles or of the continents was clearly suggested. An independent check on the palaeomagnetic determinations can be made by means of geological and palaeontological evidence for ancient climates. Thus the presence of thick coals and coral-bearing limestones in the Carboniferous and of evaporites and desert sandstones in the Permian of Europe is consistent with the palaeomagnetists' estimate of latitudes in those periods of between about 20 °N and 20 °S. Similarly the existence of late Palaeozoic glacial deposits in Australia is consistent with determination of a high palaeolatitude for that island.

Thus the new geophysical data appeared to confirm dramatically Wegener's postulation of polar wandering based on palaeoclimatic data from the stratigraphic record. Actually for some years it was not clear whether the pole moved in the same way relative to each of the continents (polar wandering) or whether different tracks were obtained from different continents, which would indicate relative movement of the continents (continental drift). The first results to suggest strongly that the continents did not all follow the same track were those of Runcorn[39] for Europe and North America. Polar wandering paths were determined for both continents but a systematic difference was discernible, the American Precambrian and Palaeozoic path being displaced about 30° of longitude to the west. After the Triassic the difference disappeared. Runcorn appreciated that the anomaly could be made to disappear if the North Atlantic were closed by bringing North America adjacent to Europe, as Wegener had proposed. The implication was that the continents had moved apart at some time between the Triassic and the present.

Although Runcorn had been trained to believe, like all other British geophysicists, that continental drift was impossible, he quickly became an enthusiastic convert and actively began to promote Wegener's basic idea. A good deal of scepticism existed about the new results but much of this was dissipated, in Great Britain at any rate, when Blackett and colleagues[40] collected the world-wide data and presented it in a way that clearly indicated the reality of drift. The results for the southern continents and India were especially convincing. In their present relative positions the polar wandering paths make no kind of sense but fall into concordance if Gondwanaland is reconstructed along the lines advocated by du Toit.

Another important research area in rock magnetism concerned so-called magnetic reversals. In the early 1950s J. Hospers at Cambridge had found many examples in young lava sequences in Iceland where the north pole flipped to the south pole and vice versa from bed to bed. Dispute persisted for a long time about whether such changes reflected reversals of the earth's magnetic field or whether they were induced in the rocks by some means related to the mineralogy. The fact that self-reversal is extremely difficult to produce experimentally, and that the direction of magnetization is the same in widely different but adjacent rock types, argues in favour of the former interpretation.

Convincing support came in the early 1960s from dating the lavas by measuring the amount of radioactive decay of potassium-40 to argon-40. It was found that rocks of the same age in widely separated parts of the world are magnetized in the same direction. Such results could not readily be dismissed as the result of random, local effects. It became apparent that the last geomagnetic field reversal had taken place one million years ago and the last previous reversal about one and one half million years earlier.

It was not immediately obvious how this could possibly relate to continental drift, but the answer soon emerged from oceanographic research.

One important reason why the pre-war debate had proved so inconclusive was that we were in almost total ignorance about what underlay the oceans which, with inland seas, cover no less than 70 per cent of the surface area of our planet. The great post-war advances in geological and geophysical oceanography have resulted from the deployment of large resources, mainly by the Americans, and development of new techniques.

Newly developed echo-sounding equipment allowed much more thorough topographic surveying than had previously been possible. Exploration through the 1950s established the existence of a world-embracing mid-oceanic ridge system, including the Mid-Atlantic Ridge and the topographically less rugged East Pacific Rise. This ridge system compares in its relative elevation with high mountain ranges on land, and is frequently displaced by what appear to be transverse faults. The crests of the Mid-Atlantic Ridge and Carlsberg Ridge (in the Indian Ocean) were shown to contain a depression that seemed to continue through the Gulf of Aden into the great East African Rift. Indeed, the American oceanographer B. C. Heezen, who claimed to trace the whole system for 35 000 miles through the ocean, interpreted the median valley as a tensional rift feature.

New methods of coring deep-sea sediment and dredging of rock samples from exposed scarps on the mid-oceanic ridges revealed the important fact that nowhere in the oceans, except on some islands of the Seychelles, were the igneous rocks of granitic type. Everywhere they consisted of basic or ultrabasic rock types, sometimes meta-morphosed, such as basalt, gabbro, amphibolite, and serpentine. Furthermore, despite extensive survey, no sedimentary rocks older than mid-Cretaceous were discovered. This fact was especially surprising, because if the oceans were the primordial features generally assumed, there ought to be some places where much older rocks than Cretaceous occurred at or near the surface.

Oceanic geophysicists studied the crustal structures by gravity and seismic methods in conjunction. Seismic refraction allowed the determination of crustal thickness and, indirectly, of density. It soon became apparent in the years after the Second World War that the ocean crust must be fundamentally different from continental crust. Both were underlain by denser material probably resembling the ultrabasic igneous rock peridotite, which was now referred to as the mantle rather than the substratum. The ocean crust was thought on strong grounds to consist of basalt and gabbro and was only about one-third the average thickness of 'granitic' continental crust. There was no indication anywhere of sunken continents.

By the end of the 1950s it was becoming increasingly apparent that the time was ripe for some new ideas. By far the most significant of these for the future history of earth science was termed by R. S. Dietz the spreading sea-floor hypothesis,[41] and was put forward by the University of Princeton geologist Harry Hess (1906–69).

Hess was that rare combination of brilliant scientist and able administrator. In the latter capacity he eventually became Chairman of the Space Science Board of the National Academy of Sciences, and was for a time the principal non-governmental adviser on the scientific objectives for planetary exploration. As a scientist he achieved great distinction as an igneous petrologist and mineralogist, with special reference to ultrabasic rocks. He provided a major stimulus to geological research in the Caribbean, and into the character of the gravity field under submarine trenches in the Pacific.

During the war, as captain of an assault transport, he directed systematic echo-sounding surveys of the Pacific, which resulted in the discovery of striking flat-topped seamounts, which he termed guyots, correctly interpreting them as deeply subsided volcanic islands. Hess' name became ultimately linked with Project Mohole, a proposal to drill beneath the ocean into the mantle, which fell foul of Congress because of escalating costs, but not before the feasibility of drilling in deep water had been established. The highly ambitious and fruitful Deep Sea Drilling Project was a natural successor.

Hess first circulated his ideas in preprint form in 1960; the published paper appeared two years later.[42] His great contribution was to integrate a number of disparate facts: the apparent youth of the ocean floor; the circum-Pacific island arc-trench system, marked by numerous volcanoes and powerful earthquakes suggestive of faults dipping to great depth away from the oceans; and the extensive mid-oceanic ridge system, with its seismicity, high heat flow, local volcanicity, and axial rift, implying tension.

He proposed that the sea floor was created at mid-ocean ridges, spread out towards the trenches and then descended beneath them into the mantle. He then related his model for sea-floor spreading to continental drift by suggesting that the continents were carried along in the process, which was driven by convection currents in the mantle, an idea long anticipated, of course, by Holmes and Fisher. In Hess's own words

a continent's leading edges are strongly deformed when they impinge upon the downward moving limbs of convecting mantle ... Rising limbs coming up under continental areas move the fragmented parts away from one another at a uniform rate so a truly median ridge forms as in the Atlantic Ocean ... The cover of oceanic sediments and the volcanic sea mounts also ride down into the jaw crusher of the descending limb, are metamorphosed, and eventually probably are welded on to continents.

A second quotation indicates the relation to drift.

The mid-ocean ridges could represent the traces of the rising limbs of convection cells while the circum-Pacific belt of deformation and volcanism represents descending limbs. The Mid-Atlantic Ridge is median because the continental areas on each side of it have moved away from it at the same rate ... This is not exactly the same as continental drift. The continents do not plough through oceanic crust impelled by unknown forces, rather they ride passively on mantle material as it comes to the surface at the crest of the ridge and then moves laterally away from it.

Sea-floor spreading confirmed

Hess's hypothesis was to be utilized in 1963 by a research student at the University of Cambridge called Fred Vine, to account for puzzling bands of magnetic anomalies occurring parallel to the Carlsberg Ridge in the Indian Ocean. Such anomalies had first been discussed and mapped in the north-east Pacific by V. Vacquier, A. Raff, and R. Mason. They occurred there as bands parallel to the crest of the East Pacific Rise and were tentatively held to delineate elongate bodies of magnetite-bearing rock, presumably basalt, with patterns of magnetization sharply contrasted with their neighbours to the east and west.

Although there was no agreed interpretation of these mysterious bands, they seemed to indicate alternating blocks of high and low susceptibility to induced magnetization. Raff and Mason[43] were sceptical of a variety of suggestions related to topographic, petrological, or thermal variations. The main interest of the Pacific anomalies at the start of the 1960s was, in fact, the evidence they provided huge lateral displacements of segments of ocean floor along fracture zones.

Shortly before Raff and Mason published their paper, magnetic anomalies on the floor of the Red Sea were interpreted as due to remanent and reversal magnetization.[44] Although the senior author, R. W. Girdler, was a pioneer in interpreting the Red Sea as a newly opening, incipient ocean tied in some way to the East African–Gulf of Aden–Indian Ocean rift system, he did not suggest global reversals in conjunction with sea-floor spreading as a means of accounting for the oceanic anomaly patterns.

This crucial insight was arrived at independently by Vine and a palaeomagnetist employed by the Canadian Geological Survey, Lawrence Morley.[45]

Vine had been sympathetic to the idea of continental drift as an undergraduate at Cambridge, and had heard Hess give a lecture there on his hypothesis in January, 1962. The head of his department,[46] Sir Edward Bullard, was a recent convert and offered encouragement, as did his supervisor, Drummond Matthews, who was also well informed about subjects pertinent to the issue in question.

It was Matthews who had proposed to Vine the detailed study of a comparatively small area on the Carlsberg Ridge in the north-west Indian Ocean. Vine found that his computer simulation model fitted the magnetic anomaly pattern well, and this led him to publish the famous article in *Nature* with Matthews.[47] As basalt was injected vertically into the axial rift zone of ocean ridges, it acquired as it cooled below the Curie Point the geomagnetic field, either normal or reversed. Subsequent injection forced the earlier-emplaced basalt to move laterally away from the ridge crest. Periodic geomagnetic reversals caused changes in 'signature' of the basalts. In Vine and Matthews' words the hypothesis

is consistent with, in fact virtually a corollary of, current ideas on ocean floor spreading and periodic reversals in the Earth's magnetic field. If the main crustal layer of the ocean crust is formed over a convective up-current in the mantle at the centre of an oceanic ridge, it will be magnetized in the current direction of the Earth's field. Assuming impermanence of the ocean floor, the whole of the oceanic crust is comparatively young, probably not older than 150 million years, and the thermoremanent component of its magnetization is therefore either essentially normal, or reversed with respect to the present field of the Earth. Thus, if spreading of the ocean floor occurs, blocks of alternately normal and reversely magnetized material would drift away from the centre of the ridge and parallel to the crest of it.

The implications of the hypothesis were extremely exciting because we had in effect a potential magnetic 'tape recorder' for determining the speed of the oceanic conveyer belt, given of course an accurate time-scale for magnetic reversals. Mapping of the anomaly patterns of the ocean floor would allow much of the most recent history of the oceans to be inferred.

In the event, despite the elegance of the hypothesis, there was a barrier of apathy or scepticism to overcome, because there was as yet insufficient supporting evidence. Matthews himself was decidedly sceptical initially, and even Vine had his apprehensions. He recalls a coffee break discussion in his department in early 1963. 'Somebody said, "Do you know if *Nature* get their articles reviewed, or do they publish almost anything?" I said, "Well we're just about to find out because, you know, I just put my paper in, and if they publish that they'll publish anything".'[48]

Well, refereed or not, *Nature* did publish, to the journal's eternal credit. Morley was less fortunate. Morley's interest in the Pacific anomalies was different from Vine's, being not so much concerned with sea-floor spreading as the application of geomagnetic reversals to geochronology. The following quotation from the letter he submitted to *Nature* in February 1963 indicates clearly, however, that he had drawn the same conclusion as Vine.

If one accepts in principle the concept of mantle convection currents rising under ocean ridges, travelling horizontally under the ocean floor and sinking at ocean troughs, one cannot escape the argument that the upwelling rock under the ocean ridges as it rises about the Curie Point geotherm, must become magnetised in the direction of the earth's field prevailing at the time.

If this portion of rock moves upward and then horizontally to make room for more upwelling material, and if, in the meantime, the earth's field has reversed, and the same process continues, it stands to reason that a linear magnetic anomaly pattern of the type observed would result.[47]

Morley's letter, upon being rejected by *Nature* in June 1963, was promptly submitted to the *Journal of Geophysical Research* but suffered a similar fate, with the referee making his now notorious comment that 'This is the sort of thing you would talk about at a cocktail party, but you would not write a letter on it'.

Shortly afterwards Morley found a journal prepared to publish his ideas, in a longer article devoted to palaeomagnetism as a means of dating geological events.[49] Thereafter his involvement with the subject ceased, and it was left to Vine to seek confirmatory evidence.

Vine was well aware that, without more concrete evidence in support, his hypothesis in some ways posed more problems than it solved, and the background assumptions about sea-floor spreading, global geomagnetic polarity reversals, and remanent magnetism were all questionable in 1963. In particular it still required to be demonstrated that the anomalies were symmetrical on each side of ocean ridges and that they were strictly parallel to their crests.

The only other research group actively exploring the ocean floor magnetic anomalies at the time was based at the Lamont Geological Observatory[50] of Columbia University, New York. This had been built up in the post-war years by its dynamic director, Maurice Ewing, into one of the greatest institutes of geological and geophysical oceanography in the world. The institute's general bias towards data-collecting on a massive scale, and hostility towards mobilistic views, reflected the outlook and attitude of its director. Of the established personnel, only Bruce Heezen and Neil Opdyke were sympathetic to continental drift. Heezen had discovered the world-wide oceanic ridge rift system and interpreted it in terms of an expanding earth, with the continental masses being dispersed as the Atlantic and Indian Oceans opened up; Opdyke, who was by now busy establishing himself as a leading authority on geomagnetic reversals as recorded in deep-sea sediment cores, had spent several years as a graduate student in Runcorn's laboratory at Newcastle-upon-Tyne.

Jim Heirtzler, Manik Talwani, and Xavier Le Pichon explored a vast area of ocean floor and recognized a remarkable symmetry in the anomaly pattern over the Reykjanes Ridge south of Iceland. Nevertheless they favoured a stabilist interpretation in articles published in 1965[51] rejecting the Vine–Matthews hypothesis, in particular because it failed to explain major differences in amplitude of the flank anomalies compared with the anomalies adjacent to the crest axis.

In that same year, however, the breakthrough came for Vine. Tuzo Wilson was, with Hess, on sabbatical leave at Cambridge and it was there that he developed his famous ideas on transform faults (see below). He and Vine decided to collaborate in a study of the Juan de Fuca Ridge off Vancouver Island, in order to see if the anomaly pattern there fitted the Vine–Matthews hypothesis; also to test the transform fault thesis. Vine and Wilson recognized symmetry in the anomaly pattern, as predicted, and attempted to estimate the rate of sea-floor spreading using the new geomagnetic reversal time-scale of the American scientists Cox, Doell, and Dalrymple. The background to this crucially important work has been explored in great scholarly detail by William Glen.[52]

The palaeomagnetists Allan Cox and Richard Doell, students of Jean Verhoogen, had conceived a research programme in the Geology Department in the University of California at Berkeley, aimed at resolving the question of magnetic reversals. Later moving to the United Geological Survey office at nearby Menlo Park, they invited Brent Dalrymple, a Berkeley rock dater, to collaborate in formulating a polarity-reversal time-scale for the last few million years; several increasingly refined versions were produced over a period of a few years. Dalrymple benefitted from the great advances made at Berkeley by Garniss Curtis and Jack Evernden, using a static-mode mass spectrometer that counted isotopes at an unprecedented level of refinement (built by John Reynolds) to date rocks of increasingly young age by the potassium-argon method. The incentive for Curtis and Evernden was improvement in the dating of early Man in East Africa; no-one at Berkeley had originally conceived using the method for the magnetic polarity time-scale. The research at the Australian National University of Ian McDougall, another Berkeley-trained rock dater, with the English palaeomagnetist Don Tarling, provided the healthy stimulus of competition for Cox and his colleagues. The eleventh version of their time-scale, containing the newly-discovered Jaramillo Event, was at Menlo Park in 1966, and quickly became the standard used by Opdyke and his colleagues in the interpretation of deep-sea cores.

The confluence of evidence for magnetic reversals, recorded by three different kinds of geological mechanisms, now fired intense interest in the Vine–Matthews hypothesis. Until the work of Cox, Doell, and Dalrymple, there had been no conclusive demonstration that rocks of identical age had the same magnetic polarity all over the earth. The co-existence of two groups of workers at Berkeley, initially with quite different research interests, appears in retrospect as a remarkable case of serendipity. By his own admission, Vine only became substantially convinced of the correctness of his hypothesis in November, at a meeting of the Geological Society of America in Kansas City when Dalrymple told him about the Jaramillo Event. Whereas the time-scale used by Vine and Wilson led to the inference of a rather irregular spreading rate which did not accord well with the magnetic anomaly pattern, the newly amended time-scale now gave a very regular and much faster spreading rate over the Juan de Fuca Ridge.

There was a striking change of view a few months later among the Lamont group, principally as a result of obtaining an outstanding magnetic anomaly profile across the Pacific–Antarctic Ridge, the so-called Eltanin-19. In the words of Walter Pitman, the person responsible and at the time a graduate student,

It hit me like a hammer... In retrospect, we were lucky to strike a place where there are no hindrances to sea-floor spreading. We don't get profiles quite that perfect from any other place. There were no irregularities to distract or deceive us. That was good, because by then people had been shot down an awful lot over sea-floor spreading. I had thought Vine and Matthews was a fairly dubious hypothesis at the time, and Fred Vine has told me he was not wholly convinced of his own theory until he saw Eltanin-19. It does grab you. It looks very much like the way a profile ought to look and never does. On the other hand, when another man here saw it his remark was 'Next thing, you'll be proving Vine and Matthews'. Actually, it was this remark that made me go back and read Vine and Matthews. We began to examine Eltanin-19, and we realised that it looked very much like a profile that Vine and Tuzo Wilson published just before our data came out of the computer.[53]

The Lamont group became finally convinced following their own independent discovery of the Jaramillo Event. Vine told them that they had been anticipated in this by Allan Cox and his colleagues, when he visited the Lamont Observatory in February, 1966. He reports a critical encounter as follows.

Walt [Pitman] and Neil [Opdyke] were analysing their data in the same room. When I walked into the room, Neil was poring over a light-table and he was drawing up the diagram – literally drawing the diagram – which appeared in his 1966 paper. Although I think the first thing we talked about was that, I distinctly remember that all Walt's profiles, the Eltanin-19 profiles, were stacked up on the opposite wall, and so I looked from one to the other. Neil said, 'Look Fred, fantastic, we just discovered a new event. We call it the such-and-such event.' I said, 'Oh yes, I hate to tell you this, Neil, but Cox, Doell and Dalrymple have discovered that event and they named it and presented it'. He was just astounded. And I said 'Yes, Neil, it is called the Jaramillo. Moreover, here it is on the Eltanin-19 profile.' They both looked at Eltanin and looked back at me. They said, 'My God!'[54]

Plate tectonics

The seminal idea of plate tectonics was put forward by the Canadian geophysicist Tuzo Wilson in a paper in *Nature* in 1965.[55] He was impressed by the fact that movements of the earth's crust were largely concentrated in three types of structural features marked by earth-quakes and volcanic activity, namely mountain ranges, including island arcs, mid-ocean ridges, and major faults with large horizontal displacement. Especially puzzling was the fact that these features frequently seemed to end abruptly along their length.

Wilson proposed that the 'mobile belts' are in fact connected by a continuous network which divides the earth's surface into several large, rigid plates. Any features at its apparent termination could be transformed into either of the other two types. Such a junction was called a *transform*. Faults on which the displacement suddenly stops or changes direction were termed *transform faults*. Wilson presented a simplified model of the opening of the Atlantic which shows how such faults develop; the offset of the Mid-Atlantic Ridge is independent of the distance through which the continents have moved and merely reflects the shape of the initial break between the continental blocks.

While Wilson was the first to use the term 'plates' the full theoret-ical formulation and development was by the Princeton geophysicist Jason Morgan.[56] (Dan McKenzie at Cambridge developed similar ideas at about the same time.)

Morgan had the idea of extending the transform fault concept to a spherical surface. He divided the earth's surface into 20 blocks, some small, some large, divided by boundaries of three types: (1) ocean rises, where new crust was created; (2) ocean trenches, where crust is destroyed; and (3) transform faults, where crust is neither created nor destroyed. In order to give the interpretative model mathematical rigour, the blocks were assumed to be perfectly rigid. The crust, especially the oceanic crust, is too thin to exhibit the required strength, and so the blocks or plates were thought to extend about 100 km down to the so-called low-velocity layer of the mantle, which makes the top of the weaker *asthenosphere*. The relatively rigid zone of the upper 100 km, termed the tectosphere by Morgan, has become more widely known as the lithosphere.

Morgan applied Euler's theorem, which states that a block on a sphere can be moved anywhere else on that sphere by a single rotation about a given axis. He was able to support calculation on the change of sea-floor spreading rate in different places by reference to Atlantic data. Le Pichon[57] recognized six major plates in a very important analysis that followed on from Morgan's work and his own consider-able knowledge of the sea-floor spreading data (there are a dozen or more minor plates such as the Carribean region). The term continen-tal drift is no longer apposite because, although the continents move, they form only part of a given plate, and certainly do not 'drift' through the oceans. A basic tenet of plate tectonics is that the amount of crust created at divergent plate margins must equal the amount destroyed by 'subduction' at convergent margins.

Plate tectonics successfully survived a number of seismological tests, based on fault-plane solutions to determine the direction of relative motion of adjacent plates. A new programme of deep-sea drilling also provided general confirmation of the accuracy of the so-called Heirtzler time-scale which was used for dating the sea floor, and which was based on the assumption that the Vine and Matthews hypothesis was true.

The late 1960s and early 1970s were a time of great excitement and activity in the earth sciences community, as new data and interpreta-tions based on plate tectonics flooded in both from oceanography and land geology. New interpretations were put forward on a wide range of major topics such as mountain building, igneous activity, sea-level changes, metallogenesis, and the biogeography of extinct organisms. The earth sciences were, by general acknowledgement, revitalized almost beyond recognition as our view of the globe has been transformed.[58]

The conversion of the earth sciences community

We have seen that in the early 1950s continental drift was taken seriously by very few. The minute number of staunch adherents tended to be dismissed as cranks, especially in the United States, a point of considerable sociological interest. A large number of people were either noncommittal or had a sneaking sympathy with the ideas of Wegener and du Toit, but considered it professionally wise to keep fairly quiet about it. Yet by the end of the 1960s the situation was completely reversed, and supporters of the discredited stabilist doctrine denying continental drift declined to a small minority of diehards.

The first cracks in the armour of the stabilist doctrine opened up in the mid-1950s with the publication of the results of rock magnetism by the Cambridge and Imperial College teams. This certainly softened resistance to the notion of mobilism in Great Britain but there is not much indication of a major change of view of the geological consensus. There was still a good deal of suspicion of the new and unfamiliar research technique and the reliability of its results were widely questioned. In North America the new palaeomagnetic work was generally received with extreme scepticism.

Another significant event of the mid-1950s was a conference on continental drift in Hobart, Tasmania, at which the guest of honour was the Yale University professor Chester Longwell, who for many years had been one of the few American geologists to go on record as being fairly agnostic to Wegener's ideas. The organizer of the conference was the Professor of Geology at Hobart, Warren Carey, who also had by far the longest and most significant article in the published proceedings.[59]

Carey is a bold and original thinker who returned to a view of tectonics on the grand scale as promoted by Suess and Argand, taking whole mountain belts as his subject matter. This was in sharp contrast to most of his structural geologist contemporaries, who were getting preoccupied with increasingly detailed and technical matters. Carey argued from his large-scale analysis that mobilism could no longer be dismissed, and also that the earth had expanded significantly in the last 200 million years; the continents had not so much drifted apart as dispersed. (This even more radical idea, while mobilist, denies the reality of the subduction process. It has attracted a small minority of supporters.)

Carey subsequently undertook extensive lecture tours in Europe and North America, forcefully presenting his views, and while his work was looked on askance by more meticulous minds there is no doubt that he influenced opinion in many circles towards a mobilist interpretation of the globe.

The distinguished English geophysicist Sir Edward Bullard was substantially persuaded by the palaeomagnetic results of the truth of continental drift by 1959, and came out in favour without reservation in a lecture to the Geological Society of London in 1963. He pointed to the increasing evidence of major horizontal displacements both on land and under the ocean, and discussed the possibility of a mantle convection mechanism along the lines of Holmes' ideas. Bullard subsequently wrote[60] that he was surprised at the amount of support shown in the subsequent discussion.

Another important London meeting took place in March 1964, when the Royal Society held a discussion on continental drift, the proceedings of which were later published as a book.[61] It was evident at this meeting that the notion of drift was largely accepted. Especially persuasive was a computer fit of the Atlantic continents by J. E. Everett and A. G. Smith, utilizing a suggestion by Bullard that Euler's theorem could be applied to this problem. The technique involved the writing of a computer program to determine the best least-squares fit of two irregular lines on a sphere. The best fit turned out to be the 1000-m contour, which approximates the true edge of the continent. The published fit is remarkably accurate and belies Jeffreys' and others' claims that Wegener's original fit involved great distortion.

We have already noted the dramatic turnabout in opinion of the Lamont Observatory. Runcorn remarked after a visit at the critical time, 'I felt like a Christian visiting Rome after the conversion of Constantine'.[62] Ewing himself remained sceptical at this early stage.

In Bullard's view the turning point for opinion in America more generally was marked by a conference in New York in November 1966 sponsored by NASA. At this the world-wide evidence from the magnetic lineations and the earthquake epicentres on the ocean ridges was presented.

The effect was striking. As we assembled on the first day, Maurice Ewing came up to me and said, I thought with some anxiety, 'You don't believe all this rubbish do you, Teddy?' At the end of the meeting I was to sum up in favour of continental movement and Gordon Macdonald against; on the last day Macdonald was unable to attend and no one else volunteered to take his place. I attempted to say what I thought Macdonald would have said, but it was unconvincing and I left it out of the published account. In my summary I pointed out how far we had gone from the traditional geological interest in the continents and their mountain systems and recommended a return to these problems. I also stressed the importance of deep-sea drilling in the verification and extension of ideas about the sea floor.[63]

By the early 1970s the conversion of the earth sciences community was virtually complete (for a retrospective attempt at quantitative documentation see Nitecki *et al*[64]). Inevitably there have remained a few dissenting voices[65] but they have not proved persuasive to the great majority. As for Sir Harold Jeffreys, he stubbornly maintained his stabilist position by resolutely discounting the new geophysical evidence, as he had earlier the geological and biological evidence, preferring to place more reliance on his estimates of the viscosity of the mantle, which seemed to him still to preclude significant lateral mobility.

A retrospective epilogue

The posthumous vindication of Wegener has naturally aroused curiosity about his scientific ability as perceived by disinterested contemporaries. A student friend in Berlin, Wilhelm Wundt, assessed him thus.

Alfred Wegener started out to tackle his scientific problems with only quite ordinary gifts in mathematics, physics and other natural sciences. He was never, throughout his life, in any way reluctant to admit that fact. He had, however, the ability to apply those gifts with great purpose and conscious aim. He had an extraordinary talent for observation and for learning what is at the same time simple and important, and what can be expected to give a result. Added to this was a rigorous logic, which enabled him to assemble rightly everything relevant to his ideas.[66]

Compare this assessment with that of Hans Benndorf, a professor of physics and colleague of Wegener at Graz.

Wegener acquired his knowledge mainly by intuitive means, never or only quite rarely by deduction from a formula, and when that was the case, it needed only to be quite simple. Also, if matters concerning physics were involved, that is, in a field distant from his own field of expertise, I was often astonished by the soundness of his judgement. With what ease he found his way through the most complicated work of the theoreticians, with what feeling for the important point! He would often after a long pause for reflection say, 'I believe such and such,' and at most times he was right, as we would establish several days later, after rigorous analysis. Wegener possessed a sense for the significant that seldom erred.[67]

The restrospective recognition that Wegener was in essentials correct leads one naturally to wonder why his drift hypothesis provoked such hostility for so long, until opposition was finally overwhelmed by a mass of new geophyiscal data. Various suggestions can be made.

1. The evidence presented by Wegener was indecisive, oceanic geology was virtually unknown, the continental fits too crude, and too few critics had personal knowledge of the highly significant late Palaeozoic strata and fossils of the southern continents.

It would be quite reasonable because of such considerations to expect a large number of earth scientists to remain somewhat sceptical or at least unconvinced, but why did not such an exciting new idea stimulate active research into testing it, rather than promoting such a vehement adverse reaction on the part of so many? Furthermore, many of Wegener's better arguments never received an adequate answer from his critics.

2. Wegener's proposed continent-moving forces were easily shown to be grossly inadequate, and there were physical difficulties associated with continents 'ploughing through the sima'.[68]

This seems a more cogent criticism until it is borne in mind that the history of science records many phenomena, such as electricity and magnetism, that were accepted on empirical grounds long before any

adequate theory emerged, while the properties of heat, light and
sound were investigated long before there was any understanding of
their true nature. A good geological example is the Pleistocene ice age,
universally accepted as fact although the causes of ice ages are still
being debated (see Chapter 4). Furthermore, Holmes' convection-
current hypothesis removed the most obvious difficulties faced by
Wegener, without apparently producing many converts to drift. Even
today, while plate tectonics is almost universally accepted the controlling forces are still poorly understood.

It is much more likely that less scientifically respectable, though
perhaps more human, reasons played a large part in the general
rejection. The game is given away by one of Wegener's strongest
critics, R. T. Chamberlin, who quoted with evident approval an
overheard remark of a participant at the New York Conference in
1926. 'If we are to believe Wegener's hypothesis we must forget
everything which has been learned over the last seventy years and
start all over again.'

Such conservative prejudice is not confined to geologists.

In nearly all matters the human mind has a strong tendency to judge in the
light of its own experience, knowledge and prejudices rather than on the
evidence presented. Thus new ideas are judged in the light of prevailing
beliefs. If the ideas are too revolutionary, that is to say, if they depart too far
from reigning theories and cannot be fitted into the current body of
knowledge, they will not be acceptable. When discoveries are made before
their time they are almost certain to be ignored or meet with opposition
which is too strong to overcome, so in most instances they might as well not
have been made.[69]

Notes

1. For historical accounts of the controversy see: Hallam, A. (1973). *A
   revolution in the earth sciences*. Oxford University Press. Marvin, U. B.
   (1973). *Continental drift: the evolution of a concept*. Smithsonian Institute
   Press. The science reporter of the New York Times, Walter Sullivan,
   gives a good popular account, with emphasis on modern developments,
   in Sullivan, W. (1974). *Continents in motion: the new earth debate*.
   McGraw-Hill, New York. See also Cox, A. (ed.) (1973). *Plate tectonics
   and geomagnetic reversals*. Freeman, San Francisco. Glen, W. (1982) *The
   road to Jaramillo*. Stanford University Press.
2. An excellent scholarly account of nineteenth century geotectonic research is given by M. Greene (1985). *Geology in the nineteenth century:
   changing views of a changing world*. Cornell University Press, Ithaca.
   Dana, J. D. (1873). *Am. J. Sci.* (ser. 3) **5**, 423: **6**, 6, 104, 161.

3. Translated into English as Suess, E. (1904–9). *The face of the earth.* Clarendon Press, Oxford.
4. Carozzi, A. V. (1970). *C. R. Séances Soc. Phys. Hist. nat. Geneva* N.S. **4**, 171.
5. Snider-Pellegrini, A. (1858). *La création et sys mystères dévoilés.* Franck & Dentu, Paris.
6. Rupke, N. A. (1970). *Nature* **277**, 349.
7. Fisher, O. (1881). *Physics of the earth's crust.* Murray, London.
8. Pickering, W. H. (1907). *J. Geol.* **15**, 23.
9. Baker, H. B. (1911). *The origin of the moon.* Detroit Free Press.
10. Loeffelholz von Colberg, C. F. (1895). *Die Drehungen der Erdkruste in geologischen Zeitraumen.* Munich.
11. Kreichgauer, D. (1902). *Die Äquätorfrage in der Geologie.* Steyl, Berlin.
12. Ampferer, O. (1906). *Jahrb. K. K. Geol. Reichsanst.* 56, 539.
13. Taylor, F. B. (1910). *Bull. Geol. Soc. Am.* **21**, 179.
14. It has generally been assumed that Wegener was unaware of Taylor's work and developed his ideas independently. This has been thrown into some doubt by the recent discovery of a letter by Taylor, written in 1931. In it he expressed confidence that Wegener was indeed well aware of his hypothesis but never acknowledged it (Totten, S. M. (1980) *Geol. Soc. Am. Abstra.* p. 536).
15. The fullest biographical account is by his widow, Wegener, Else (1960). *Alfred Wegener.* Brockhaus, Wiesbaden. A more recent German book is by Schwarzbach, M. (1980). *Alfred Wegener und die Drift der Kontinente.* Wissenchaftliche Verlagsgesellschaft, Stuttgart. This has now been translated into English as *Alfred Wegener – the father of continental drift* by Science Tech., Inc., Madison, Wisconsin (1986). A brief English language biographical memoir is given by his fellow Greenland expeditioner-Johannes Georgi in *Continental drift* (ed. S. K. Runcorn). Academic Press, London (1962). See also Hallam, A. (1975). *Sci. Amer.* **232** (2), 88.
16. Wegener, A. (1966). *The origin of continents and oceans.* Translated from the 4th revised German edition of 1929 by J. Biram. Methuen, London. Unless otherwise specified, all the cited Wegener quotations come from this book.
17. See Chapter 5.
18. Lake, P. (1922). *Geol. Mag.* **59**, 338.
19. Jeffreys, H. (1926). *The earth, its origin, history and physical constitution.* Cambridge University Press.
20. Jeffreys, H. *op. cit.* (Note 19), p. 261.
21. van Waterschoot van der Gracht, W. A. J. M. (ed.) (1928). *Theory of Continental drift: a symposium.* Am. Assoc. Petrol. Geol.
22. Coleman, A. P. (1925). *Nature* **115**, 612.
23. Gregory, J. W. (1929). *Quart, J. Geol. Soc. Lond.* **85**, lxviii.
24. My italics.
25. Ich gläube, Du hältst meinen Urkontinent fur phantastischer, als er ist, und siehst noch nicht, dass es sich lediglich um Deutung des

Beobachtungsmaterials handelt. Wenn ich auch nur durch die Überein-
stimmung der Kustenkonturen darauf bin, so muss die Beweisfuhrung
naturlich von den Beobachtungsergebnissen der Geologie ausgehen.
Hier werden wir gezwungen eine Landverbindung zum Beispiel zwis-
chen Südamerika und Afrika anzunehmen, welche zu einer bestimmten
Zeit abbrach. Den Vorgang kann man sich auf zweierlei Weise vorstel-
len: Erstens durch Versinken eines verbindenden Kontinents, oder,
zweitens durch Auseinanderziehen von einer grossen Bruchspalte.
Bisher hat man, von der unbewiesenen Vorstellung der unveränderliche
Lage jeden Landes ausgehend, immer nur das erste beruscksichtigt und
das zweite ignoriert. Dabei widerstreitet das erste aber der modernen
Lehre der Isostasie und überhaupt unseren physikalischen Vostel-
lungen. Ein Kontinent kann nicht versinken, denn er ist leichter als das,
worauf er schwimmt. Also ziehen wir doch einmal das weite in Betracht!
Und wenn sich hier nun eine solch Fülle überraschender Verein-
fachungen ergibt, wenn es sich zeicht, dass jetzt Sinn und Verstand in
die ganze geologische Entwicklungsgeschichte der Erde kommt, warum
sollen wir zogern, die alte Anschauung uber Bord zu werfen? Ist dies
revolutionär? Ich glaube nicht, dass die alten Vorstellungen noch zehn
Jahre zu leben haben. Noch ist die Lehre von der Isostasie nicht
vollständig durchgearbeitet, sobald sie es sein wird, wird der Wider-
spruch zutage liegen und die alten Vorstellungen werden korrigiert
werden. (E. Wegener, *op. cit.* (Note 15), p. 75.

26. Der Brief von Professor P. ist typisch! Er will sich nicht belehren lassen.
    Die leute, die so darauf pochen, auf dem Boden der Tatsachen zu
    stehen und mit Hypothesen durchaus nichts zu tun haben wollen, sitzen
    doch allemal selbst mit einer falschen Hypothese drin!... Aus seinem
    Brief spricht nirgends das Bestreben, den Dingen wirklich auf den
    Grund zu kommen, sondern nur die Freude am Herumreiten auf den
    Unvollkommengheiten anderer Menschen. Solch Menschen sind für
    eine Neuorientierung von Ideen nicht zu haben. Hatten sie die Ver-
    chiebungstheorie schon auf der Schule gelernt, so wurden sie sie mit
    demselben Unverstand in allen, auch den unrichtigen Einselheiten ihr
    ganzes Leben hindurch bertreten, wie jetzt das Absinken von Kontinen-
    ten (E. Wegener, *op. cit.* (Note 15), p. 162.

27. Gestern habe ich anlasslich der Tagung der "Internationalen Erdmes-
    sung, mit dem Geodäten Helmert, Albrecht and Förster über Alfreds
    Kontinentalhypothese gesprochen. Es hat mich gefreut, wie ernst sie sie
    nehmen. Namentlich der alte Albrecht bezeichnete sie al sehr interes-
    sante Idee, die durchaus gepruft werden musse. (E. Wegener, *op. cit.*
    (Note 15), p. 77).

28. Ich stehe noch ganz unter dem Eindruck Ihres glänzenden Vortrags.
    Gerade deshalb, weil er so knapp war und sich nur auf die wichtigsten
    Tatsachen beschrankte, wirkte er so uberwältigend. Sie hatten mir zwar
    gesagt, ich lasse mich leicht überzeugen. Mir solchem Tatsachenmate-
    rial, durchwebt von einer zusammenfassenden Idee, schon! Dass sich
    alle geologishen Details in das Ihnen entworfen Bild nicht ohne weiteres

hineinfügen lassen, stört mich nicht in mindesten. Ebensowenig jeden, der sich mit der Erforschung des Mechanismus komplexer Naturs- cheinungen befasst hat. Denjenigen hingegen, welche ihr ganzes leben nur Tatsachen gesammelt und aufgezeichnet haben, mangelt die Fähig- keit, hinter diese Tatsachen tiefer zu blicken, ihr blick ist zue sehr auf die Oberfläche geheftet. Dass Sie exakte Wissenschaftler leichter zu über- zeugen vermogen als die Empiriker, das soll Sie nicht entmutigen, im Gegenteil. (E. Wegener, *op. cit.* (Note 15)A, p. 163).

29. Argand, E. (1924). *C. R. XIII Cong. Géol. Int.* Liege, p. 171.
30. Daly, R. A. (1926). *Our mobile earth.* Scribner, New York.
31. For a detailed account of Holmes' contribution to drift theory, see Frankel, H. (1978). *Brit. J. Hist. Sci.* **11**, 130.
32. Holmes, A. (1931). *Trans. Geol. Soc. Glasgow* **18**, 559.
32. du Toit, A. L. (1937). *Our wandering continents.* Oliver & Boyd, Edinburgh.
34. Simpson, G. G. (1943). *Am. J. Sci.* **241**, 1.
35. Simpson was eventually converted to a belief in continental drift by the evidence of oceanic geophysics, and H. Frankel has argued that, left to their own evidence, palaeobiogeographers might still be arguing the issue (in *Studies in history and philosophy of science* Vol. 12, pp. 211–60 (1981)). On the other hand Cenozoic mammals did not provide a particularly good test case.
36. Willis, B. (1943). *Am. J. Sci.* **242**, 509. Suess, incidentally, was not German but Austrian.
37. Gevers, T. W. (1950). *Trans. Geol. Soc. S. Afr.* **52** (suppl.), 1.
38. Koenigsberger, J. K. (1938). *Terr. Magn. Atmos. Elec.* **43**, 119, 299.
39. Runcorn, S. K. (1956). *Proc. Geol. Assoc. Canada* **8**, 77.
40. Blackett, P. M. S., Clegg, J. A. and Stubbs, P. H. S. (1960). *Proc. R. Soc. London* **A256**, 291.
41. For a detailed account of Hess's development of the sea-floor spreading hypothesis see Frankel, H. (1980). In *Scientific discovery: case studies* (ed. T. Nickles), p. 345. Reidel, New York. There has been much dispute about whether it was Hess or Dietz who first conceived the idea, as opposed to the name, of sea-floor spreading. H. W. Menard, whose engagingly written book *The ocean of truth: a personal history of global tectonics* (Princeton University Press, 1986) gives an authoritative account of the oceanographic research undertaken in the 1950s, which provided a necessary basis for the subsequent advances, argues for a genuine example of multiple discovery.
42. Hess, H. H. (1962). In *Petrologic studies, a volume to honour A. F. Buddington*, ed. A. E. Engel, H. L. James, and B. F. Leonard, p. 599. G.S.A. Boulder, Colorado.
43. Raff, A. and Mason, R. (1961). *Bull. Geol. Soc. Am.* **72**, 1268. Menard (*op. cit.*, Note 41) was a researcher at the Scripps Institute during this period; he did important work on the Pacific fracture zones. His insights into why the Scripps researchers 'missed the boat' makes fascinating reading. Thus Vacquier's speculations on the possible origin of the

magnetic stripes on the Pacific floor strongly hinted at the Vine and Matthews interpretation, but when confronted with their idea he rejected it.

44. Girdler, R. W. and Peter, G. (1960). *Geophys, Prosp.* **8**, 474.
45. A detailed account of the development, reception, and acceptance of what he calls the Vine–Matthews–Morley hypothesis is given by H. Frankel, in *Historical studies in the physical sciences* **13**, 1–39 (1982). See also Glen (1982) *op. cit.* (Note 1).
46. Then the Department of Geodesy and Geophysics, now the Bullard Laboratories, Department of Earth Science, University of Cambridge.
47. Vine, F. J. and Matthews, D. H. (1963). *Nature* **199**, 947.
48. Quoted in Frankel *op. cit.* (Note 45).
49. Morley, L. W. and Larochelle, A. (1964). *Roy. Soc. Canada Spec. Publ.* no. 8, p. 39.
50. Now renamed the Lamont–Doherty Geological Observatory.
51. See especially Talwani, M., Le Pichon, X., and Heirtzler, J. R. (1965). *Science* **150**, 1109. This is the last stabilist interpretation of the East Pacific Rise anomalies before general acceptance of sea-floor spreading.
52. Glen, W., *op. cit.* (Note 1).
53. Quoted in Wertenbaker, W. (1974). *The floor of the ocean*, p. 203. Little and Brown, Boston.
54. Quoted in Frankel *op. cit.* (Note 45).
55. Wilson, J. T. (1965). *Nature* **207**, 343.
56. Morgan, W. J. (1968). *J. Geophys. Res.* **73**, 1959.
57. Le Pichon, X. (1968). *J. Geophys, Res.* **73**, 3661.
58. See, e.g. Hallam (1973, *op. cit.*, Note 1) or indeed any modern general geological textbook.
59. Carey, S. W. (1958). Geol. Dept. Univ. Tasmania, Sympos. No. 5, p. 177.
60. Bullard, E. C. (1975). *Ann. Rev. Earth Planet Sci.* **3**, 1.
61. Blackett, P. M. S., Bullard, E. D., and Runcorn, S. K. (eds.) (1965). *A symposium on continental drift*. Royal Society, London.
62. Quoted in Frankel *op. cit.* (Note 45).
63. Bullard *op. cit.* (Note 60). Menard (*op. cit.*, Note 41) argues that Ewing was rather more open-minded in the early 1960s about sea-floor spreading than has been generally believed.
64. Nitecki, M. H., Lemke, J. L., Pullman, B. W., and Johnson, M. E. (1978). *Geology*, **6**, 661.
65. See, for example, articles in Kuhle, C. F. (ed.) (1976). *Am. Ass Petrol Geol. Mem.* **23**.
66. Quoted in Georgi *op. cit.* (Note 15).
67. Wegener gewann seine Erkenntnisse meist durch instinktartige, innere Anschauung, nie oder nur ganz selten durch Deduktion aus einer Formel, die aber dann auch nur ganz einfach sein durfte. Auch wenn es sich um physikalische Fragen handelte, die seinem eigenen Arbeitsgebiet ferne lagen, staunte ich oft über sein sichere Urteil. Mit welcher Leichtigkeit fand er durch die verwickelsten Arbeiten der Theoreitker

durch, mit welchem Spursinn für das Wesentliche. Er pflegte dann oft nach einer ziemlich langen Pause des Nachdenkens zu sagen: 'Ich glaube, die Sache ist so', und meistens war sie auch so, was wir oft erst nach mehreren Tagen erkennen konnten, nachdem der Beweis gelungen war. Wegener besass einen selten trugenden Sinn fur Wirkliche. (E. Wegener, *op. cit.*, Note 15, p. 179).

68. Jeffreys and his fellow geophysicists were undoubtedly much impressed by the seismic evidence that the earth responded to major earthquake shocks like a huge bell; in other words it had considerable strength. The height of mountains appeared to corroborate this, otherwise they should surely have collapsed. On the other hand, besides the varied geological evidence for drift marshalled by Wegener and his supporters, there was the clear evidence from the hearts of mountain belts that tough metamorphic rocks that shatter under hammer blows were intensely contorted in a way implying plastic deformation. The resolution of what Oxburgh (1981, *Nature* **293**, 26) has called the strength paradox comes in an understanding of the phenomenon of *creep*, whereby materials subjected to relatively low stresses for very long periods of time deform continuously like highly viscous fluids. Needless to say, Wegener appreciated the point fully, as expressed in the analogy he drew with pitch.

69. Beveridge, W. I. B. (1950). *The art of scientific investigation*. Heinemann, London.

7
Mass extinctions

As we have seen in Chapter 2, Cuvier was the first person to recognize that certain organisms, such as the mammoth and the giant sloth, had become extinct. He went on to argue for episodes of mass extinction and in effect became the father of catastrophist thought in the early nineteenth century. Cuvier was a greatly respected figure in France and acquired a number of disciples, one of whom, d'Orbigny, introduced the concept of stage into stratigraphy on the basis of extinctions of successive faunas (see Chapter 3). Another disciple, Elie de Beaumont, put forward a widely accepted theory which related mass extinction events to the environmental disturbance provoked by sudden episodes of mountain upheaval. Under the influence of Lyell, however, English geologists tended to reject Cuvier's catastrophism so that Darwin's mid-century view, expressed in the *Origin of species*, that dramatic faunal turnovers in the stratigraphic record reflected major hiatuses, provoked no dissent. 'The old notion of all the inhabitants of the earth having been swept away by catastrophes at successive periods is very generally given up.' Darwin, indeed, was firmly of the opinion that biotic interactions, such as competition for food and space, were of considerably more importance in promoting evolution and extinction than changes in the physical environment.

Species are produced and exterminated by slowly acting causes ... and the most important of all causes of organic change is one which is almost independent of altered ... physical conditions, namely the mutual relation of organism to organism – the improvement of one organism entailing the improvement or the extermination of others.

Thus the marked differences between Palaeozoic, Mesozoic, and Cainozoic faunas, recognized by John Phillips as early as 1840, did not elicit much interest in terms of extinction, and the subject was largely ignored until early this century, when T. C. Chamberlin proposed that major faunal changes through time, which provided the basis for stratigraphic correlation, were under the ultimate control of epeirogenic movements of the continents and ocean basins. Marine regressions corresponded with periods of continental rejuvenation and ocean basin subsidence and led to provincialism and areal restrictions on organism distributions, while transgressions corresponded with periods of reduction of the continents to base level and allowed the spread of new faunas.[1] Chamberlin's was an isolated article that did not stimulate a major research programme, and indeed the study of extinction events remained dormant for many years. However, his influence can be discerned in important papers by distinguished American palaeontologists written half a century later. The first of these, by Raymond C. Moore,[2] concerned itself with extinctions of North American late Palaeozoic marine faunas, which were attributed to crowding effects produced by the areal restriction of epicontinental seas at times of regression.

Of more general significance was a series of papers by Norman Newell of Columbia University, culminating in one pertinently entitled 'Revolutions in the history of life'.[3] Besides the major extinction events at the end of the Palaeozoic and Mesozoic eras, Newell recognized four other events in the Phanerozoic which were also dramatically sudden in the context of the much longer time intervals preceding and following them, and which affected a significant proportion of the world's marine organisms. These took place at the end of the Cambrian, Ordovician, Permian, Triassic, Cretaceous, and near the end of the Devonian; all but the end-Cambrian event have become generally accepted subsequently by palaeontologists as being by far the most important mass extinction events in the stratigraphic record. Newell related extinctions to regressions, as did Moore, and considered these to be global in extent, and produced a similar argument for the cause of extinction of neritic organisms, namely restriction of habitat leading to a deterioration in environmental conditions. Radiation of the survivors took place during the expansion of habitat area consequent upon a succeeding transgression. Although attempts were made to quantify the phenomenon that Newell described in terms of what ecologists call the species–area effect, it has to be admitted that extinctions were not a major preoccupation of the new wave of palaeobiologists who began in the 1970s to revitalize the study of fossils from an evolutionary point of view. The stimulus for the new interest in mass extinctions that has dominated palaeontological research in the 1980s came from an altogether unexpected source.

In the middle of this century the German palaeontologist Otto Schindewolf, who had long been preoccupied with the marine mass extinction at the end of the Palaeozoic, came to the conclusion that the event must have been a catastrophic one for which he could literally conceive no earthly explanation. Consequently he was led to speculate that the extinctions had been caused by a near-by supernova explosion. The increased cosmic radiation impinging on the earth had destroyed the ozone shield and led to lethal exposure of numerous organisms.[4] A similar idea to account for mass extinctions was put forward later by a physics professor at Columbia University, Malvin Ruderman, and by some Russian scientists. In 1970 the expatriate British palaeontologist, Digby McLaren, who had risen to become Director of the Canadian Geological Survey, gave his presidential address to the Palaeontological Society.[5] As with Schindewolf, he considered that the marine mass extinction at the end of the Frasnian stage (late Devonian) was much too widespread, dramatic and 'geologically instantaneous' to be caused by a merely terrestrial process, and speculated that the world's ocean had been severely disturbed by a giant meteorite impact. Three years later the American chemist and Nobel Laureate Harold Urey had a paper published in *Nature* which argued for several extinction events within the last 50 million years being caused by the impact of comets.

What these various suggestions had in common, with a few others invoking increases of radiation from outer space, either cosmic radiation or solar protons, is that they were virtually ignored. This is unsurprising in view of the almost total absence of supporting evidence, excluding perhaps a few tektite layers in Tertiary deposits. The situation was to change dramatically in 1980 with the publication of an article by a group of scientists at Berkeley, California, including the physicist and Nobel Laureate Luis Alvarez, his geologist son Walter and the nuclear chemists, Frank Asaro and Helen Michel.[6]

The asteroid impact hypothesis[7]

According to Walter Alvarez the origins of the impact hypothesis can be traced back to the beginning of the 1970s, when he began a collaboration at the Lamont-Doherty Geological Observatory with the Scottish geophysicist Bill Lowrie on the palaeomagnetism of pelagic limestones in Italy. Close to the pretty mediaeval town of Gubbio, in the heart of the Umbrian Apennines, is a beautifully exposed section in the Bottacione Gorge of a complete Upper Cretaceous and Lower Tertiary sequence. These limestones exhibit an excellent record of magnetic reversals, which could be correlated with the biostratigraphic zonation based on planktonic foramimifera. It had been discovered by the micropalaeontologists who worked on the section in the 1960s that there was a distinctive 1 cm clay bed exactly at the Cretaceous–Tertiary boundary, which was marked by the massive disappearance of nearly all the forams, with only one of the smallest species continuing into the Tertiary. Research in other parts of the world had shown that none of the disappeared species reappeared in the Tertiary elsewhere, so the event in question must mark a mass extinction.

A few years later Alvarez invited the Princeton geologist Al Fischer to give a lecture at the Lamont about the events at the end of the Cretaceous, and was reminded of something he had virtually forgotten, that the mass extinction event affected not only the planktonic forams (and a group of nannoplankton also with calcium carbonate skeletons, the coccolithophorids) but also such familiar groups as ammonites and dinosaurs. Fischer stressed that the cause of the end-Cretaceous extinction was unknown, and Alvarez was prompted to think that the boundary clay layer at Gubbio might possibly yield some evidence that had a bearing on the subject. Therefore in the summer of 1977 he carefully collected some samples for analysis.

When he joined the faculty at the University of California at Berkeley that autumn, he began to discuss the extinction problem with his father, who had been well established there as a particle physicist for many years. Recently retired, his ever-curious mind was constantly on the lookout for new problems to tackle, and the long-standing enigma of dinosaur extinctions posed a great challenge. One obvious question to attempt to resolve was how long it had taken to deposit the clay layer. Luis Alvarez's suggestion was to measure its iridium content, and that of the strata immediately above and below. Whereas iridium, which belongs to the so-called platinum group of metals, is severely depleted in the earth's crustal rocks, it is relatively enriched in meteorites, especially iron meteorites (though still only present in trace quantities expressed as parts per billion). The Alvarez' assumption was that iridium from meteoritic dust accumulates at a more or less constant rate in sediments and thus will make a useful geological timer. (Iridium had already been used by geologists at the Scripps Institute of Oceanography to determine the cosmic content of Pacific floor sediments.) The iridium content can be determined precisely, to parts per trillion, by neutron activation analysis. Luis Alvarez' colleagues at the Lawrence Berkeley Laboratory, Frank Asaro and Helen Michel, are established experts with this technique, and were therefore invited in 1978 to collaborate in the research project they devised.

To everyone's surprise it turned out that there was far more iridium in the Gubbio boundary clay than could be accounted for by the normal rain of micrometeorites. In the summer of 1979, while Walter Alvarez was in Italy doing fieldwork, his father set himself the task of finding an explanation for the huge iridium enrichment. He continually bombarded Asaro and Michel with ideas, most of which he rejected himself, and for a time the group got excited by the possibility that the anomalous iridium could be attributed to a nearby supernova explosion. However, Asaro and Michel showed that the platinum 244 isotope that ought to characterize a supernova deposit was not present, so a different hypothesis was required.

Further work had shown that the iridium anomaly was essentially confined to the thin boundary clay, and the iridium content quickly declined to normal background levels above and below in the stratal section. Comparable anomalies were found in Cretaceous–Tertiary (or K–T) boundary sections, as determined from micro- and nanno-plankton, in both Denmark and New Zealand, strongly suggesting a global rather than a local event. Luis Alvarez eventually hit upon the idea of impact by an asteroid with the composition of a chondritic meteorite. This would have expelled a huge amount of pulverized crustal rock, laced with iridium–rich asteroid material, to create a dust cloud that would have enveloped our planet for up to a year. Such a cloud would have blocked out sunlight, thereby stopping photosynthesis and resulting in a collapse of the food chain, with consequent starvation and mass extinction. The boundary clay at Gubbio and elsewhere marked the material that had eventually settled from the globally-distributed dust cloud. On this assumption, and knowing the elemental composition of chondritic meteorites, Alvarez calculated an approximate diameter for the asteroid of 10 km, which would imply the creation of an immense impact crater of the order of 150–200 km in diameter. The impact hypothesis was first brought into the public domain as a preliminary report at a meeting of the American Association for the Advancement of Science in San Francisco, and was published in an article in *Science* in May 1980.[8]

Early reactions to the hypothesis

The University of Chicago palaeontologist Dave Raup was one of the referees of the Berkeley group's article. As he reports in his book entitled *The nemesis affair*[9] he was sympathetic but sceptical. 'If a graduate student gave me this manuscript to read, I would see it as a brilliant piece of work (indicating that the student had enormous potential) but I would give it back to be done right.' According to Raup most of the comments he heard from geologists and palaeontologists during the months following publication were dismissive towards the hypothesis, for a variety of reasons. There was no evidence that terrestrial and marine extinctions were strictly contemporary. Too little was known about iridium geochemistry. It could have been concentrated by organisms in the sea, or selectively adsorbed by organic matter on the sea floor. Better stratigraphical control was needed – iridium anomalies might prove to be common in the geological column once they were looked for, and not associated with mass extinction events. Where was the huge crater demanded by the hypothesis?

The palaeontologists' reaction was epitomized by an article entitled 'Out with a whimper not a bang', published a year later by three Americans, Bill Clemens, Dave Archibald, and Leo Hickey.[10] The first two were mammal specialists but with a good knowledge also of reptiles, the last a palaeobotanist. According to this trio the fossil record in the Western Interior of the United States, which for the K–T non-marine boundary beds is the best in the world, showed that the extinctions were not catastrophic but gradual and selective in a way incompatible with the drastic environmental scenario put forward by the Berkeley group. Indeed, many groups of organisms, such as freshwater reptiles, had evidently survived virtually unscathed. A complex interaction of physical and biological factors was invoked instead to account the increase in extinction rate at the end of the Cretaceous. While their views were widely shared in the palaeontological community, three prominent impact sympathizers emerged. Not surprisingly for one who had a decade earlier proposed a meteorite impact hypothesis himself, Digby McLaren quickly became an enthusiastic supporter and has continued to campaign with great eloquence on behalf of the catastrophic nature of mass extinctions, which to him demands an extraterrestrial explanation. The Canadian Dale Russell, alone among dinosaur experts, maintained that the end-Cretaceous dinosaur extinctions were indeed catastrophic and were not preceded, as widely believed, by a period of decline lasting for many thousands or even a few million years. The Dutch micropalaeontologist Jan Smit had been working on a K–T boundary section in pelagic limestones and marls at Caravaca, in southern Spain. He quickly became a convert to the Alvarez hypothesis, to the extent of publishing a paper in *Nature* advocating an asteroid or comet impact shortly before the Berkeley group's article appeared, though their assistance was duly acknowledged. Smit had learned about the iridium anomaly at Gubbio at a meeting the previous year and has persuaded a geochemist colleague to analyse his samples of the boundary clay at Caravaca. Once again, a striking anomalous enrichment of both iridium and its sister metal, osmium, was found exactly at the level which marked an abrupt extinction of planktonic forams and coccoliths; the preceding deposits gave no hint of any decline in these groups.

In the same issue of *Nature* there appeared a paper by the cosmopolitan geologist Ken Hsü, who was born in China, had a long spell in the United States, and is now a Swiss citizen.[12] He had also learned about the Gubbio anomaly, and was stimulated to come up with a hypothesis of cometary impact for the end-Cretaceous extinctions. Large terrestrial animals were killed off by intense heating of the atmosphere while the calcareous plankton were destroyed as a consequence of cyanide poisoning and a catastrophic rise of the calcium carbonate compensation depth in the oceans, following cyanide detoxification. Although he has subsequently abandoned this particular idea Hsü has continued to argue with his own characteristic blend of enthusiasm and engaging candour the case for catastrophism in geology – often extraterrestrially induced – and especially as a cause of mass extinctions, railing with equal vehemence against the uniformitarians Lyell and Darwin.[13]

The idea of atmospheric heating as a means of killing off the dinosaurs had been proposed a few years earlier by the American palaeontologist Dewey McLean, who noted that temperature change can disturb the sex ratio in living reptiles. If the only reptiles to hatch are *either* males *or* females the group is unlikely to survive for long! Atmospheric heating as a killing mechanism was also proposed by Cesare Emiliani, Eric Kraus, and Gene Shoemaker. They were stimulated by the Alvarez hypothesis to write a paper proposing an impact in the ocean.[14] Initially global temperatures would decline drastically because of the opaque blanket of atmospheric dust, but would then rise substantially as a consequence of the greenhouse effect induced by massive amounts of water vapour.

The Berkeley group's findings stimulated a great deal of geochemical and mineralogical research, and before long many other K–T boundary horizons in widely-distributed marine deposits revealed an iridium anomaly. Particularly notable was the work of Ramachandran Ganapathy, employed by a chemical company in New Jersey; he realized that other rare siderophile (or 'iron-loving') elements, such as platinum and osmium, and related elements (such as gold) should behave in the same manner as iridium, and that ratios of rare siderophiles in the boundary clay could be used to 'fingerprint' the chemistry of the asteroid.[15] He found that the ratios of these elements from the boundary clay at Stevns Klint, Denmark, matched rather closely those of chondritic meteorites, something that was confirmed for other K–T boundary localities by Asaro and Michel, and by Frank Kyte and his colleagues at the University of California at Los Angeles. Another important discovery was made by Carl Orth, of the Los Angeles National Laboratory, and colleagues in his laboratory and the United States Geological Survey.[16] They found an iridium anomaly in a non-marine section in New Mexico, precisely at the K–T boundary as defined by pollen grans. This confirmed a prediction of the impact hypothesis, that the anomaly should be global, occurring on both land and under the sea, and appeared to rule out decisively the alternative hypothesis, put forward by Michael Rampino, that the iridium was extracted from sea water. Not long afterwards the Berkeley group found another iridium anomaly in a non-marine K–T boundary section in Montana.

Jan Smit was the first to recognize crystalline spherules, a few tenths of a millimetre in diameter, in the boundary clay at Caravaca. Such spherules have subsequently been found in the boundary clay at other sites, including Gubbio. They have been interpreted as the cooled droplets of impact melt – in other words microtektites – altered after formation into minerals such as sanidine feldspar, glauconite and magnetite, but retaining features in some instances interpreted as relict quench textures, with precursor minerals such as plagioclase feldspar and pyroxene.

Of a series of conferences at which the impact hypothesis and its consequences were discussed, none was more important than that held at the ski resort of Snowbird, Utah, in the autumn of 1981, and the published proceedings give a good measure of the 'state of the art' at that time.[17] It would not be simplifying matters unduly to state that the conference participants divided rather clearly into impact enthusiasts and sceptics, the former group being dominated by physical scientists and the latter by palaeontologists. Participants at the meeting recall that many discussions were flavoured with no small amount of acrimony.

Three of the most interesting papers presented at the meeting concern the drastic changes to the environment that would have been induced by impact. John O'Keefe and Tom Ahrens of the California Institute of Technology inferred that vaporized, melted, and small solid ejecta would transfer nearly half their energy to the atmosphere, giving rise to a brief and possibly lethal heating pulse, followed by global cooling due to the reduction of light levels by dust and vapour. Like the model of Emiliani and his colleagues, this involved spectacular changes in global atmospheric temperature, but in reverse order. The model put forward by Brian Toon and his colleagues suggested that large quantities of dust would remain in the atmosphere for only three to six months before settling. Whereas the oceans would cool by only a few degrees because of their large heat capacity, continental temperatures would drop well below freezing for approximately twice as long as light levels insufficient for photosynthesis persisted, which would kill all animals that were unable to hibernate. A group of atmospheric chemists from the Massachusetts Institute of Technology

introduced the concept of acid rain on a massive scale, arguing that asteroid or comet impacts must produce large quantities of nitrogen oxides. Rapid production of these oxides and subsequent 'rainout' of nitric and other strong acids would acidify the surface waters of the ocean, selectively destroying calcium carbonate skeletons. This could help to account for the selectivity of the extinctions, with calcareous plankton being the most severely affected.

While the geochemists present were for the most part impact supporters, Frank Kyte and Jean Wasson of the University of California at Los Angeles introduced a cautionary note. The pattern of siderophile elements, though generally chondritic in character, varied from site to site at the K–T boundary, and there must be a considerable continental source for the boundary clay; it could not all be impact-fallout material. Other geochemists argued that certain isotope ratios signified an oceanic impact, but existing data allowed the alternative of an authigenic origin and were therefore ambiguous. Richard Grieve, a cratering expert with the Canadian Geological Survey, reviewed the known impact structures. Approximately a hundred such structures ranging in diameter up to 140 km had been recognized on the continents. Large oceanic impacts would bring upper mantle material to the earth's surface and produce detectable geophysical anomalies but none had yet been detected. The record of terrestrial cratering was not inconsistent with the end-Cretaceous impact proposed, but it does not supply any direct confirmation of a relationship with mass extinctions.

As regards the palaeontologists, Norman Newell, the doyen of students of mass extinctions, expressed what could be termed the standard view, that the extinctions were not catastrophic but took place over significant time intervals lasting as long as several million years. The normal method of reporting the temporal ranges of fossils artificially concentrates the last occurrence at stratigraphic boundaries. Furthermore, unrecognized so-called paraconformities, which signify gaps in the stratigraphic record without any clear physical evidence, might simulate mass extinction events. Helen Tappan likewise argued that the timing and selectivity of most extinctions were not compatible with sudden events. Concerning the stratigraphic sections in Denmark that had begun to figure so largely in the K–T boundary dispute, Tove Birkelund reported that the fossil data pointed to a prolonged biotic turnover throughout the latest Maastrichtian (the youngest Cretaceous stage) and the early Danian (the oldest Tertiary stage). This signified not a single catastrophe but a more complex environmental scenario. Tom Schopf maintained that the story of the final demise of the dinosaurs in effect resolved itself into what happened to less than twenty species in the Western Interior of North America. As the interior seaway regressed, temperature seasonality increased and the habitable area was reduced as widespread floodplains yielded to uplifted restricted basins. The gradually changing ecology implied by these geological events provided a sufficiently plausible scenario for dinosaur extinctions without the need to invoke an impact-induced catastrophe in addition.

The calcareous plankton not unnaturally received considerable attention since it was these organisms that appeared to record a catastrophic event most obviously. Hans Thierstein confirmed that the end-Cretaceous extinction event did indeed appear to be 'geologically instantaneous', but his Swiss compatriot Katerina Perch-Nielsen noted that relict nannoplankton became extinct several thousand years after the end of the Cretaceous, suggesting a longer-term environmental stress rather than a catastrophic change. On the other hand Jan Smit continued to argue for a short-term catastrophe, with estimates of sedimentation rate leading to the suggestion of mass extinction within 50 years, and a new stable planktonic fauna being established within 35 000 years.

The most original and enduring paper to emerge from the palaeontologists was, however, devoted to methodology, and its implications were far from hostile to impact supporters. Phil Signor and Jere Lipps, of the University of California at Davis, pointed out that the apparent decline of taxa prior to mass extinctions might simply reflect sampling effects and not actual diversity trends. Thus the appearance of a gradual extinction can be reasonably generated by sampling effects alone even when the extinction is actually sudden. The sampling problem is minimal for calcareous micro- and nannoplankton, because of the huge number of specimens that occur at closely spaced intervals, and it is these that demonstrate the best evidence for inferring catastrophic change. It is maximal for large, comparatively rare organisms with limited preservation potential, such as dinosaurs.

Subsequent research in support of the impact hypothesis

By 1983 research in several laboratories in North America and Europe had established about 50 K–T boundary iridium anomalies across the world, including several in continental sections in the United States Western Interior. These latter anomalies coincided closely with a drastic reduction in the proportion of angiosperm pollen at the expense of fern spores, giving rise to the so-called fern spike. This implied a brief episode of ecological trauma for the flowering plants, with the ferns representing the first colonizers after the disaster; it appeared to call into question Hickey's anti-catastrophist claims for the plant record. At the Snowbird Conference the Yale University geochemist Karl Turekian had proposed an ingenious test that he believed would show the K–T boundary clay to be of terrestrial rather than extraterrestrial origin. This was based on the isotopic ratio of osmium 187 to osmium 186, which is now about unity for meteorites, but between 13 and 30 for crustal rocks. The measurements he made later with his colleague Jean-Marc Luck showed that the K–T boundary clay osmium was in fact definitely not crustal but could be of meteorite origin. A mantle origin could not, however, be excluded.

Research in palaeomagnetism appeared to confirm that the plankton and dinosaurs had indeed gone extinct during the same magnetic zone, 29 R, and some confusion about the zone recording dinosaur extinctions in the United States Western Interior was cleared up. Unfortunately this only gives a time resolution of about half a million years, so the exact contemporaneity required by the impact hypothesis was not established. Japanese scientists studied a section of marine deposits in Hokkaido and found a fern spike in the palynological record coincident with the K–T boundary as determined from planktonic foraminifera. This appeared to strengthen the link between marine and continental extinctions.

The most significant new discovery, however, was made by Bruce Bohor, of the United States Geological Society at Denver. He found grains of so-called shocked quartz in the same clay in the United States Western Interior that contained the iridium anomaly.[18] Further grains were found more rarely at a few other sites in Europe. Shocked quartz possesses distinctive lamellae which have only been recognized in rocks from well established meteorite impact sites or at nuclear weapons test sites, and appears to signify the passage of shock waves under enormous pressures. Needless to say, this discovery proved immensely exciting to impact supporters because it seemed to provide the independent confirmation they needed, and many erstwhile fence-sitters became converted.

Another new technique involved analysing the boundary clays rich in organic matter at the now-classic localities of Stevns Klint, Denmark, Caravaca, Spain, and Woodside Creek, New Zealand. Electron microscope examination of material separated from the dominant kerogen showed the samples to contain up to half a percent of graphitic carbon as fluffy aggregates which were interpreted as soot.[19] Because these localities are widely scattered, and on the assumption that the Alvarez hypothesis was correct – with the boundary clay being laid down in at most a year – a worldwide layer of soot was inferred, which could greatly have enhanced the darkening and cooling of the earth by rock dust. The surprisingly large amount of soot inferred from this bold extrapolation, equivalent to some 10 per cent of the present biomass of the earth, would imply either that much of the earth's vegetation was burned down in a series of wildfires on an immense scale, or that much fossil fuel was also ignited. The devastating environmental scenario so envisaged would surely have given rise to massive extinctions of at least terrestrial organisms.

Geological criticisms

The sceptical response of many palaeontologists to the impact hypothesis has already been noted, but a number of criticisms were also made on geological grounds. The absence of the required huge crater of the right age was readily conceded by many impact supporters to be a problem, because a large number of older craters have been preserved in the geological record. An impact site on the ocean floor would, of course, be much more difficult to detect than one on the continents, and indeed one postulated locality of the appropriate shape and size was in the Amirante Basin in the Western Indian Ocean. Unfortunately the discovery of shocked quartz, while supportive of impact by an extraterrestrial body such as an asteroid or comet (grouped together as *bolide*) posed an embarrassment for proponents of oceanic impact, because it was difficult to see where all the quartz would have come from. Others pointed out that the clay mineral composition of the K–T boundary clay varied from site to site, which suggested local derivation by normal geological processes rather than geologically instantaneous fallout from a global dust cloud.

The most comprehensively argued critique came, however, from two marine geophysicists at Dartmouth College, New Hampshire, Charles Officer and Charles Drake – both usually known by the uniquely American nickname Chuck. Officer, in particular, was to prove the most doggedly determined opponent of K–T impact. In the first of their two articles in *Science*[20] several major points were made. Examination of K–T boundary rock in both onshore sections and Deep Sea Drilling Project cores revealed a range of transition times for the faunal turnover, which was not always intimately associated with an iridium anomaly. This anomaly forms a sharp spike only in some sections. Others showed a vertical distribution that could not adequately be accounted for by disturbance by burrowing organisms. Variations in noble metal abundances in relation to chondritic meteorites indicates some fractionation of the siderophile elements, and the role of seafloor or within-sediment reducing conditions in controlling the chemistry had been wrongly discounted or under-estimated. The evidence does not support a globally distributed boundary clay, because many sections lacked such a clay. The end of the Cretaceous was a time of great geological disturbances including volcanism, climate, and oceanography, any of which might have played a role in the extinctions, and were not geologically instanta-neous but lasted over a period of time up to about a hundred thousand years or longer.

In their second *Science* article published two years later[21] Officer and Drake took into account a significant new discovery made on the Kilauea volcano in the Hawaiian islands. Airborne particles from a recent eruption had been found to be enriched in iridium by 10 to 20 000 times the concentration in normal Hawaiian basalt. At 600 parts per billion the iridium concentration in these particles compared with that in meteorites. Officer and Drake therefore put forward an alternative to impact to account for the iridium enrichment, that it was derived from the mantle over a prolonged period of between about ten and a hundred thousand years in association with eruption of the so-called Deccan Traps. These are flood basalts occupying a vast area of India, known to be of the right age and already implicated in the end-Cretaceous extinctions by McLean.

Officer and Drake enlarged on the iridium enrichments in some sections, which were distributed over a portion of section implying sedimentation over several thousand years. Where sharp iridium spikes occurred, they were associated with lithological change, sug-gesting that concentrations might be bound up with sedimentary environmental conditions. The microtektite interpretation of bound-ary clay sperules was challenged, and it was pointed out that coesite, a silica polymorph indicative of extremely high pressures, had been found in association with kimberlites and eclogites, undoubted mantle rocks. Thus the so-called shocked quartz might also have a mantle source, and be associated with 'cryptoexplosion' features such as the Vredefort Dome in South Africa, a Precambrian structure that has also been interpreted as being a consequence of bolide impact.

In the following years the Dartmouth geophysicists teamed up with others in an attempt to substantiate a terrestrial alternative to impact. In a letter provoked by the soot article by the University of Chicago chemists Wolbach, Lewis and Anders, Officer and Ekdale[22] pointed out that their work did not provide independent support for impact but depended wholly on the correctness of the original Alvarez hypothesis. If the boundary clay were deposited at the same rate as the rock of similar lithology above and below, rather than within a year, then the concentration of organic matter at the boundary would only be between three and four times higher, which is nothing exceptional. Furthermore the section with the best data, Stevns Klint, is condensed relative to other sections, implying either a very low sedimentation rate or subsequent dissolution of calcium carbonate. The grandiose extrapolation to a global wildfire was entirely un-justified in the circumstances.

Neville Carter, of Texas A and M University, one of the world's leading experts on shock deformation of minerals, was brought in to collaborate on the possible relationship of this to explosive volcanism, with reference specifically to lavas of the famous Toba eruption in Indonesia and the Bishop Tuff of California. This led to a sharp exchange of views with the US Geological Survey group in Denver, who had discovered shocked quartz in the Western Interior, and among whom Glen Izett had come to emerge as a significant figure.[23] Despite long experience of volcanic rocks Izett had failed to find any evidence of lamellar features of the sort found in the boundary clay, and concluded that only the compressional pressures consequent upon impact could account for them.

The Canadian geochemist Jim Crocket was also involved in a group effort to reanalyse the classic Gubbio section, leading to the con-clusion that anomalous iridium, though it certainly peaked exactly at the K–T boundary, was distributed over a thickness of strata indicative of many thousands of years of deposition. I accepted an offer from Officer to join his group in a comprehensive treatment of the end-Cretaceous extinction story, my contribution being to review the biological data.[24] The conclusion reached was that the extinctions were complex and selective, taking place over an extended period of time, and therefore not plausibly accounted for by a single impact. Extinction rates in many groups increased as a result of marine regression and climatic deterioration, and received a catastrophic *coup de grâce* at the end of the period, signalled by the iridium anomaly and shocked quartz; this was thought more likely to be due to massive volcanism than impact. The obvious source would be the Deccan, especially as a team of French scientists led by Vincent Courtillot had recently confirmed that the basalts were erupted over a period of up to about a million years within magnetic zone 29 R, that is the horizon which embraces the K–T boundary. It had become well known that aerosols carried into the stratosphere from such flood basalts could produce acid rain on a considerable scale over a large area of the planet, provided volcanism was sustained and intensive enough, just as dust from explosive volcanoes could cause a lowering of temperature through restriction of sunlight.

This series of papers served to focus attention on the duration of the end-Cretaceous extinction event. If the extinction took place within a year then it was certainly catastrophic by any normal use of the term. If it extended over many thousands of years another less dramatic term would be called for, and the possibility of a whole complex of interacting environmental factors would need to be looked into very closely. Unfortunately the possibility of analysing the stratigraphic record to the degree of temporal refinement required to resolve the issue is normally very limited.

Reactions to criticism by the Berkeley group

Luis Alvarez had the opportunity of replying to his palaeontological critics in a lecture subsequently published by the National Academy of Sciences in 1983.[25] Most of this article was devoted to a review of the evidence supporting the impact theory and discussing the most likely environmental scenario for mass extinction. That a 10 km-diameter asteroid should have hit the earth within the last 100 million years was entirely plausible, as Gene Shoemaker had demonstrated by invoking the power law relationship between size and frequency of impact, which had been verified by lunar studies.

The argument with the palaeontologists focused on a long-standing debate between Alvarez and his Berkeley colleague, Bill Clemens, concerning the significance of the position of the highest dinosaur remains in the classic Hell Creek section in Montana, which occurred three metres below the iridium anomaly. Alvarez had persistently tried without success to persuade Clemens that his table showing the distribution of dinosaur bones was consistent with there having been no change prior to the end of the Cretaceous. Much of the argument hinged on sampling statistics, with a 3m gap not being held to be significant for such large, and hence relatively uncommon, organisms; his frustration at what he evidently interpreted as obtuseness is apparent from the following quotation.

I really cannot conceal my amazement that some palaeontologists prefer to think that the dinosaurs, which had survived all sorts of severe environmental changes and flourished for 140 million years, would suddenly, and for no specific reason, disappear from the face of the earth ... in a period measured in terms of thousands of years.[26]

The physicist Alvarez clearly had little patience with the different approach to solving scientific problems entertained by palaeontologists!

When leading fossil experts persisted nonetheless in disagreeing with Alvarez, he said, with some exasperation, 'I'm really quite puzzled [that] knowledgeable palaeontologists would show such a lack of appreciation for the scientific method', and, giving his side of a dispute with two palaeontologists about the time when the last dinosaurs vanished, 'I'm really sorry to have spent so much time on something the physicists in the audience will say is obvious'.[27]

Alvarez and his colleagues were no less sparing with the two geophysicists who had challenged their story in the prestigious journal *Science*. Officer and Drake were taken to task for various sins of omission and commission and accused of applying a double standard, being hypercritical of evidence supporting impact and uncritical of evidence against it. The concluding statement by Alvarez *et al.* is patronizing.

The last paragraph of Officer and Drake's article seems to be a plea for a return to the time before the iridium anomaly was discussed, when almost any speculation on the K–T extinction was acceptable. This idea is pleasantly nostalgic, but there is by now a large amount of detailed astronomical, geological, palaeontological, chemical, and physical information which supports the impact theory. Much interesting work remains to be done in order to understand the evolutionary consequences of the impact on different biological groups, but the time for unbridled speculation is now past.[28]

Nevertheless, despite their strong response to Officer and Drake's 1983 paper, Alvarez and his colleagues made an important concession. A period of 10 to 100 000 years was not an unreasonably long time for an extinction episode caused by environmental deterioration triggered initially by impact. This was evidently an attempt to meet criticisms from many geological and palaeontological sceptics, but marked a notable backtracking from their original position. Such a revised interpretation would, of course, imply that nothing adverse had happened to the biosphere until the bolide struck, at the very end of the Cretaceous. However, in another paper published in *Science* the same year, a different conclusion was drawn.[29] The Berkeley group had teamed up with two well-known scientists who had expertise in both palaeontology and sedimentology, the American, Erle Kauffman, and the Dane, Finn Surlyk. While some extinctions of marine benthic organisms appeared (from the Danish sections at least) to have taken place abruptly at the K–T boundary, contemporary with the plankton extinctions, others, such as the rudist bivalves, a dominant reef-forming group in the Cretaceous tropics, preceded the boundary by a geologically significant period of time, probably as much as several million years. The end of the Cretaceous was in fact preceded by a period of increased extinction rate in at least some marine organisms. The obvious implication was that the end-Cretaceous mass extinction in the marine realm was to be seen as a complex, drawn-out affair that could not plausibly be attributed to a single impact event. The contrast with the triumphant tone of Luis Alvarez's National Academy article could not have been more marked, but by no means did it mean abandoning some form of impact scenario, because Kauffman and Walter Alvarez were later to collaborate with others to put forward a hypothesis of multiple impact to account for the end-Cretaceous extinctions.

Extinction periodicity[30]

Jack Sepkoski of the University of Chicago published in 1982 a compendium of Phanerozoic marine vertebrate, invertebrate, and protist families known from the fossil record, with their stratigraphic ranges given to the nearest stage. Both he and his colleague, Dave Raup, were renowned in the palaeontological community for their sophistication in statistical matters, and in 1983 they got to work analysing the compendium. Using a combination of methods involving Fourier analysis and Monte Carlo simulation, they discovered to their great excitement a statistically highly significant 26-million-year periodicity of extinction events within the last 250 million years. This was a most surprising result, because rare events such as floods and hurricanes are apparently randomly distributed in time. In their paper published early in 1984[31] they concluded that 'it seems inescapable that the post-late Permian extinction record contains a 26-Myr periodicity'. Since they could not conceive of any terrestrial process that would be so periodic they were inclined to favour an extraterrestrial cause. For Raup it was a remarkable *volte face* for someone who had established a reputation for scepticism and the use of the null hypothesis.

Well before the paper was published many people knew about their astonishing findings. Sepkoski had presented the results at an extinctions symposium at Flagstaff, Arizona, the previous autumn, preprints had been circulated and the media aroused, with articles appearing in both the scientific and popular press. Thus it was that only two months after the Raup and Sepkoski paper appeared in print, the *Nature* issue for April 19th contained five papers that were written in direct response to it, a phenomenon that led to the editor complaining in that issue about the circulation of preprints to a select inner circle, cutting out others who were not in the network. (In his book, Raup exculpates himself from this venal sin by pointing out that the scientists in question had *requested* preprints, having heard of the periodicity work through articles in the press.)

The five papers fell into three categories. Walter Alvarez and his Berkeley astrophysicist colleague Richard Muller had teamed up to subject the record of impact craters on the earth to time series analysis. Only 13 craters, of a much larger number of known craters, were adequately dated, but these were claimed to be sufficient to enable a 28 million year periodicity to be detected. Then there were two astronomical hypotheses relating extinction events to comet impact, the galactic plane and the companion star, each put forward independently by two groups of researchers.

The galactic plane hypothesis, proposed independently by the American astronomers Michael Rampino and Richard Stothers, and by Richard Schwartz and Philip James, which owes a debt to the pioneer work in this field of the British astronomers Victor Clube and Bill Napier,[32] takes account of the fact that the sun crosses the plane of our galaxy twice in each cycle lasting between about 62 and 67 million years. Close to the plane there is an increased likelihood of encountering giant clouds of gas and dust, the so-called molecular clouds. These could perturb the cloud of numerous comets close to the solar system known as the Oort cloud – widely accepted by astronomers though never observed – causing a few of these comets to hit the earth, perhaps several in a period of time up to about a million years, with disastrous consequences for many organisms. Thus periodic extinctions could perhaps be explained, though the period would be nearer to 30 than 26 million years.

The alternative hypothesis, put forward independently by Marc Davis, Piet Hut, and Richard Muller, and by Daniel Whitmire and Albert Jackson, postulated that the sun had an unseen companion star, subsequently dubbed Nemesis after the Ancient Greek goddess who ensured that no mere mortal ever challenged the dominance of the gods. This star had a highly eccentric orbit. When near the perihelion it was supposed to perturb the orbit of comets in the Oort cloud, thereby initiating an intense comet shower upon our planet. Thus the end result from the earth's point of view was much the same, whichever hypothesis might be preferred.

Until this time I had been an interested bystander concerning the extinctions controversy, but I began to be drawn into it by accepting an editorial invitation to contribute a *News and Views* piece for the *Nature* issue, commenting on the five papers and the article that had started it all. My general reaction was one of scepticism. I questioned the reliability of the geological time-scale utilized by Raup and

Sepkoski. Because of the poor quality of many of the data on which it was based, and the large errors involved, use of other time-scales might have found no periodicity. I also noted of their extinction events that some were extremely dubious, and others were minor, and deviant from the purported periodicity by several million years. The argument for cratering periodicity through the Phanerozoic, based on so few data points, was not readily believable. What I reacted to most strongly was that a group of astronomers seemed to be blithely entering the extinctions debate without making any attempt to learn more about what geology had to say on the subject, for instance about global changes in sea level and climate. 'Before astronomers indulge in further speculations about the cause of mass extinctions they would do well to learn something about the rich stratigraphic record of their own planet.' One was left with the impression of a science in which too many theoreticians were chasing too few facts, a situation very different from that in geology.

Two different controversies were generated as a result of the Raup and Sepkoski article and its immediate aftermath. The first of these concerned the rival merits of the astronomical hypotheses. The Nemesis hypothesis was criticized because, firstly, the companion star has never been observed (and still hasn't despite several years of intensive searching by Muller) and secondly, its orbit would be unstable because of the gravitational deflection induced by passing stars; thus there could be no periodicity. The galactic plane hypothesis is less easy to dismiss, and indeed the well-established periodicity is tantalizingly close to that claimed by Raup and Sepkoski. However, it faces the serious problem that the sun is currently very close to the galactic plane but the last extinction event was 11 million years ago; according to the hypothesis the sun should thus be at the maximum distance away. A group of British astronomers have pointed out another serious if not insuperable difficulty with the hypothesis: the molecular clouds required for cometary perturbation are too sparsely distributed to make encounters with the Oort cloud plausible.[34]

The second controversy concerned the extinction periodicity. Raup and Sepkoski countered my criticism of use of a particular time-scale by repeating their analysis with other time-scales, and showing that the periodicity persisted, although the error margins were greater. Considerable discussion was provoked by the Polish palaeontologist, Antoni Hoffman, who argued in a paper in *Nature* that the apparent periodicity could have resulted from purely random processes. Raup and Sepkoski's claims were strongly dependent on a series of arbitrary decisions concerning the dating of stratigraphic stage boundaries, the culling of the database, and definition of mass extinction events. The average duration of a stage is 6 million years, and 4 times 6 is close to 26 million years. Applying a strict random walk model (0.5 probability increase and decrease of the extinction metric regardless of what happened during the preceding step) there is a 0.25 probability of any particular stage representing an extinction peak. Thus such peaks are to be expected at approximately every fourth stage.

An exciting result based on statistics is bound to attract statisticians into the fray, and two University of Chicago colleagues of Raup and Sepkoski duly obliged. Stigler and Wagner observed that a major component of their extinctions periodicity analysis was a significance test decisively rejecting the alternative hypothesis that extinctions occurred randomly. Stigler and Wagner confirmed this result, but discussed two things that led them to conclude that the apparent periodicity could be a statistical artefact. Certain types of measurement error can enhance a periodic signal or cause a pseudoperiodic signal to emerge from aperiodic data. The 'hypothesis of a periodic dynamic structure is so powerful in its implications, and so selective in the ease with which it imposes itself on us with limited data sets such as this one, that it must be required to pass a stringent test'.[36]

Criticism of a quite different sort, namely the taxonomy utilized, came from two palaeontologists at the Natural History Museum in London, Colin Patterson and Andrew Smith.[37] Both of them staunchly committed to the cladistic method of phylogenetic classification of organisms, they examined the data for the two groups of which they had specialist knowledge, the fish and echinoderms (a significant proportion of the total data) and found them wanting. Only about one-quarter of the purported family extinctions were real in the sense of signifying the disappearance of monophyletic groups. The remainder is mere 'noise', being the result of 'extinctions' of non-monophyletic groups, mistaken dating and 'families' containing only one species.

These criticisms represent only the most discussed of the many provoked by Raup and Sepkoski, with a number of statements in support.[38] In reply to Hoffman, they rejected his statistical argument, claiming that it reflects an incorrect application of their neutral model. When correctly applied it actually supports periodicity rather better than their original analysis! In replies to the criticisms of Stigler and Wagner, and Patterson and Smith,[38,39] use was made of Sepkoski's new generic dataset, which supports the analysis at family level, and genera are better than families as a proxy for species, the only real biological entities. Most of the extinctions emerging from the family- and generic-level analyses were well recognized by palaeontologists, and included such major events as those at the end of the Permian,

Triassic, and Cretaceous. They certainly could not be dismissed as taxonomic artefacts.

Despite the widespread scepticism about the underlying astronomical hypotheses, and the extreme paucity of evidence that could be used in support, multiple cometary impacts as a cause of mass extinctions were promoted in an article in *Nature*[40] by a group of eight authors. These included, besides those two renowned impact supporters Walter Alvarez and Gene Shoemaker, the Princeton astrophysicist Piet Hut (one of the proponents of the Nemesis hypothesis) and four palaeontologists, Erle Kauffman, William Elder, Gerta Keller, and Thor Hansen. The article can be clearly seen as an outcome of the collaboration between Alvarez and Kauffman initiated several years earlier, which had led to joint publication in *Science*. It posed the question: if at least some mass extinctions are caused by impacts, why do they extend over intervals of several million years, and have a partly stepped character? Three extinction events were cited, the Cenomanian–Turonian boundary (in the mid-Cretaceous), the end-Cretaceous boundary, and the Eocene–Oligocene boundary. Judging from reactions expressed at international conferences within the following two years the article by Hut and his colleagues met with general scepticism. The only persuasive physical or chemical evidence for impact was at the K–T boundary (and that was for a single impact) and for the later Eocene (with a couple of tektite horizons). The latter, however, did not appear to correlate with mass extinction events. Sampling problems make it doubtful if stepped extinctions can be reliably distinguished from either catastrophic or gradual extinctions. Despite the abundant evidence from the stratigraphic record of sea-level regression, climatic deterioration, and oceanic anoxia at the time intervals cited, the authors chose to disregard these more plausible, earth-bound causes as possible mass extinction promoters.

Assessment of the controversy in the late 1980s

Early in 1987 the editors of *Science* commissioned two review articles on the K–T boundary mass extinctions, which were meant to appear in the same issue, one arguing the case for impact, the other the case against. Luis Alvarez was invited to write the first article but eventually declined because he had just written a similar article for *Physics Today*.[41] Thus my article arguing in favour of an entirely earth-bound environmental scenario appeared on its own in November that year.[42]

Alvarez' article exhibited his usual lucid style, with various points argued logically and with characteristic trenchancy, but did not differ in any substantial way from his 1983 article. As in that article he expressed the case for bolide impact as a series of predictions for which supporting evidence was subsequently found. Shocked quartz had not been discovered until after he had written his earlier article, and was not predicted at that time. Needless to say it could have been, and Alvarez exploited his opportunity, treating it effectively as the final conclusive support required. Scorn was poured on the volcanic alternative to account for the iridium anomaly and shocked minerals, but judgement was reserved on cometary showers.

My article summarized the principal criticisms that had been levelled at the impact hypothesis, and argued the case for a compound extinction scenario involving marine regression and climatic fluctuations with a final *coup de grâce* being delivered by volcanism on a spectacular scale. Note was taken of recent work which suggested that the boundary microspherules might have had an algal rather than a tektite origin, and of a recent theory that related crustal movements and volcanism of different types to phases of disturbed activity in the mantle.

In October 1988 a second conference at Snowbird, Utah, was organized by the Lunar and Planetary Institute, entitled Global Catastrophes in Earth History, and devoted to the themes of impacts, volcanism, and mass mortality.[43] Like its predecessor it was interdisciplinary to a degree unusual in science, with specialists from disciplines ranging from astrophysics and atmospheric chemistry to different branches of geology and palaeontology. Of all the many conferences on extinction held in recent years this one was perhaps the best to give a measure of what consensus might have been achieved after eight years, and of relevant research in progress.

Most attention was, unsurprisingly, devoted to the K–T boundary. There appeared to be a consensus in favour of impact rather than volcanism to account for the shocked quartz, and general acceptance of Glen Izett's invocation of the 30 km-diameter Manson crater in Iowa as the likely impact site in North America, the only region where shocked minerals are anything but extremely rare. Any relationship of this relatively modest impact event to global extinctions was, however, unclear. A case was made for an impact-related tsunami event in the North Atlantic, based primarily on complex sedimentary structures found in a section in the Brazos River, Texas. This interpretation provoked much discussion, but no agreement on a possible impact site; it could hardly have been Iowa!

In contrast to shocked mineral grains the iridium story was seen as complex. Earlier attempts to locate the approximate site of the purported impact by tracing concentration gradients geographically had proved unsuccessful, and it became evident that secondary factors associated with normal precipitation from seawater had complicated the geochemistry of the boundary clay. Nevertheless something like 100 sites exhibiting the anomaly had by now been found, a high proportion of all those examined. Frank Asaro thought the evidence favoured only one anomaly but many disturbed K–T boundary sections. Walter Alvarez recognized that the Manson crater was inadequate to account for the global iridium anomaly and associated extinction events, and speculated that there might have been several simultaneous impacts as a consequence of a comet breaking up near the earth. Birger Schmitz argued that the boundary clay was locally derived and could not be simple fall-out material. Whereas he found the iridium anomaly at Stevns Klint to be associated with organic matter, Gunter Graup found it to be associated with sulphide in coal in a Bavarian section; there were three distinct iridium spikes there, the biggest being 16 cm below the K–T boundary as determined by micropalaeontologists. Attention is turning increasingly to the ratio of iridium to other related elements in the hope of improving the distinctiveness of the geochemical fingerprint. Thus it has been argued by impact supporters that the iridium/gold ratio of the Kilauea iridium-rich volcanic material rules out a mantle source. This, however, overlooks the facts that Kilauea was only one small eruption, and that we know immensely less about mantle chemistry as a whole than about meteorite chemistry.

Unlike at the first Snowbird conference, a number of palaeontologists presented an abundance of pertinent data, which appeared to impress the physical scientists present. The excellent terrestrial plant record in North America clearly indicates a complex pattern of change which was not marked by major extinction, and not confined to the K–T boundary. The dinosaur story remains tantalizing. The only good record comes from a limited part of North America and even here, despite intensive search, there is still a widespread 2m gap between the highest recorded fossil and the iridium anomaly. The latest Cretaceous dinosaurs probably coexisted with mammals hitherto thought to be Palaeocene; there was little support for Palaeocene dinosaurs, but this was perceived as crucial. As regards the marine plankton, two of those best qualified to judge, Gerta Keller and Jan Smit, failed to agree whether the extinctions were truly catastrophic (Smit) or extended over thousands of years (Keller). Peter Ward's impressive new data from the Bay of Biscay coast suggested that a number of ammonite genera had survived to the very end of the Cretaceous, thereby contradicting earlier claims, and demonstrating how careful and thorough collecting can alter our perception of extinction events.

With regard to possible extinction scenarios there was nothing remotely approaching consensus. The trouble with the various impact-related scenarios that have been put forward, whether involving spectacular changes of temperature, acid rain, or wildfires, is that they are too drastic to account for the selectivity of the extinctions, with many groups of varied biology passing across the K–T boundary more or less unscathed. Thus Ronald Prinn, an atmospheric chemist at the Massachusetts Institute of Technology, presented an update of the acid rain scenario and argued that it was an effective way to account for the mass extinction of calcareous plankton, by lowering the alkalinity of surface ocean waters and dissolving calcite skeletons. However, the ocean is far more buffered against changes in alkalinity than most lakes, yet many freshwater organisms, most notably the crocodiles, were virtually unaffected by the extinction event, and there is no evidence of lakes in limestone country, that might have been so buffered, in the North American stratigraphic record. Edward Anders, an expert on the chemistry of meteorites at the University of Chicago, fiercely defended the wildfire hypothesis against numerous critics, but the relentless combativeness he displayed throughout the meeting was uncharacteristic of the general tone, which was notable for its lack of acrimony despite an abundance of vigorously expressed disagreement. Environmental factors such as sea-level and climatic change that could have had a longer-term influence, and thereby helped to explain better those extinctions not concentrated exactly at the K–T boundary, received remarkably little consideration.

Other extinction events were only briefly considered. Richard Muller's attempt to relate numerous cometary impacts to glacio-eustatic sea-level changes and magnetic reversals was coolly received, as was Michael Rampino's idea that massive and long-sustained volcanism could be triggered by impact. Carl Orth reported that his quest for iridium anomalies at extinction boundaries apart from the K–T had yielded negative results, except for modest anomalies associated with the Cenomanian–Turonian boundary event, and these were more likely to be related to events in the mantle than to impact. The leading tektite expert, Billy Glass, of the University of Delaware, perceived little relationship between Eocene tektite-strewn fields and mass extinctions of marine micro-organisms.

One of the most striking features of the aftermath of the Alvarez impact hypothesis for the end-Cretaceous extinctions is the way controversy has cascaded into at least eight disputes, none of which can be said to be totally resolved as yet, in spite of a continually increasing amount of research. These can be expressed most vividly as a series of questions.

1. Iridium anomaly. Is this single or multiple? Is it related to impact or volcanism? The role of organic matter, sea water, sediment condensation, and sedimentary sulphide have not yet been entirely eliminated from consideration because they appear to play a major role, at least in the level of concentration.

2. Shocked minerals. Are they related to impact or volcanism? Though a consensus appears to exist in favour of impact, high pressure polymorphs of silica such as coesite are known from mantle-derived rocks and a rare, unusually explosive event unrelated to impact cannot yet be completely ruled out.

3. Microspherules. Are these altered from microtektites or algal spheres? Need they all have a common origin – could some be normal micrometeorite rain concentrated by condensation, others volcanic?

4. Boundary clay. Is this impact fall-out from a global dust cloud, or locally derived by normal sedimentational processes?

5. Impact event. Was this single or multiple, oceanic or continental, or both? Comet or asteroid? If multiple, were the impacts simultaneous or not?

6. Deccan traps vulcanicity. Was this triggered by impact or not?[44]

7. Extinctions. Were they catastrophic, gradual, or stepped, or a mixture of all of these?

8. Killing scenario. Was the sunlight blotted out by a dust cloud? Did the earth's surface temperature rise or fall? Was there a global wildfire, or acid rain on a massive scale? Not all these postulates are, of course, mutually exclusive. What was the role of a sharp sea-level fall and subsequent rise, for which there is strong evidence from the stratigraphic record? It has always seemed a remarkable coincidence that the purported bolide struck precisely at the time when one of the world's more significant eustatic changes took place, and only Muller has suggested that large-scale sea-level change could be triggered by impact.

If we extend consideration to mass extinctions in general, then two other continuing controversies must be added. Are the extinction events genuinely periodic or merely episodic? If one is to believe in the periodic impact of showers of comets, which hypothesis is the more plausible, the galactic plane or Nemesis?

General evaluation

Whatever the truth may prove to be, there can be no doubt that the Alvarez impact hypothesis has been of immense heuristic value. It has focused attention in an unprecedented way on a major scientific problem, and stimulated a considerable amount of fruitful research. Wider ramifications that have been explored as a direct consequence include nuclear winter scenarios and astronomical factors, such as impact by cometary showers that visit the earth at periodic intervals. Lively interactions between scientists of disciplines as varied as astrophysics, chemistry, and palaeontology have resulted in a plethora of new data and interpretive models. These interactions have brought out rather dramatically the very different approaches to their subject adopted by physical scientists on the one hand and geologists and palaeontologists on the other. The former group tends to concentrate on producing testable predictive models which inevitably oversimplify Nature, with full use being made of probability assessments; they are deeply suspicious of data that cannot meaningfully be quantified. The latter group, in contrast, feel that they cannot avoid engaging in data gathering, with verbal and diagrammatic description on a massive scale, if they are to do justice to the complexity of the natural world. Interpretation is integral to their subject matter but tends to be qualitative in essence. Analytical skills are more prized among the former, synthetic skills among the latter.

Leading impact supporters within the geological community, such as Walter Alvarez, Ken Hsü and Gene Shoemaker, have frequently stated that the conservative reaction of many of their colleagues has been the consequence of indoctrination in their student days with Lyellian uniformitarianism, such that catastrophism became a dirty word. It is at least as plausible to argue that the reaction is a natural response of resentment by scientists trained in the classical disciplines of historical geology and palaeontology at the brash entry into their subject area by physicists and chemists. Some of these scientists, they would maintain, have displayed a combination of ignorance and naïvety, occasionally tainted with arrogance, towards the manifest complexities of the stratigraphic record, which can only be deciphered patiently as a result of intense effort in a variety of research fields. There are no 'magic bullets' that will solve major problems in a trice, and results derived from sophisticated 'black box' technology are often as ambiguous and indecisive as those derived from simpler tools.

It is, however, not difficult to find grounds for optimism. Throughout the debate engendered between impact supporters and their opponents, there have been accommodations on both sides, as was apparent at the second Snowbird conference. Palaeontologists have been stirred out of their previous complacency about extinctions to collect data in order to test specific predictions, and seem more ready to accept at least some degree of catastrophic change. Impact

supporters are more willing than hitherto to acknowledge the extended time intervals through which at least some extinctions took
place, the phenomenon of extinction selectivity, and the complications
introduced into simple model building by climatic and oceanographic
change, episodic volcanism, and tectonism. Research groups involving the collaboration of palaeontologists, sedimentologists, and
geochemists have targeted particular extinction events, and fruitful
interaction between model builders and data gatherers offers much
promise for the future.

In recent years historians of science have increasingly focused
attention on the importance of social and psychological factors. The
mass extinctions controversy offers a rich field of investigation, which
can only be briefly touched on here. Tales are told of extravagant
efforts of self-promotion by some individuals in the mad scramble to
climb on the bandwagon and obtain research grants, of less than
scrupulous attempts to discredit opponents, and even of pressure to
block the attainment of tenure by aspirant junior academics who held
the 'wrong' opinions. Whatever the truth of such allegations, and only
the most naïve would deny that such a competitive field of science has
its darker side, it would be wrong to be unduly influenced by them
into thinking that good science has been the loser, because there has
in fact been an abundance of this, with most of its practitioners
exhibiting the more decent human qualities.

In 1984 an interesting opinion poll was conducted on the response
by a sample of various groups of scientists to the Alvarez impact
hypothesis. Four statements were put forward for agreement or
disagreement by the different groups, which included American
palaeontologists and geophysicists and British palaeontologists; the
results are as follows.[45]

	American geophysicists	American palaeontologists	British palaeontologists
1. An extraterrestrial impact at the K–T boundary caused mass extinction	31	16	9
2. There was a K–T impact but other factors caused mass extinction	15	39	27
3. There was no K–T impact	13	6	23
4. There was no K–T mass extinction	3	8	13

There is clearly a gradient of increasing scepticism towards the Alvarez hypothesis from American geophysicists to British palaeontologists (Polish palaeontologists were even more sceptical). This is likely to be a reflection of the more conservative attitudes of palaeontologists than geophysicists, as discussed earlier. Another factor could be that British (and other European) palaeontologists are even more conservative and may register a greater reluctance to be swayed by the latest scientific fashion or, less charitably, have a poorer acquaintance with the relevant scientific literature.

That the impact theory was fashionable can be discerned from any number of articles in the popular and popular scientific press, and on television and radio programmes. To the general public it would appear that a combination of dinosaur extinctions and death stars is irresistible. The role of the media in the extinctions debate has been more significant than in any other scientific controversy.[46] Its influence has been bad insofar as facts and opinions have been oversimplified, distorted, or sensationalized, but good where it has disseminated knowledge quickly to a wide audience, facilitating a transfer of information between different disciplines. Those two flagships of responsible science journalism, *Science* and *Nature*, have shown an active involvement from the start, and have published a disproportionately high number of the key papers in the dispute. Whereas *Science* has been accused by some of showing in its commentaries an uncritical bias towards impact, the editor of *Nature* has twice chosen the subject as a theme for critical articles. Firstly he berated a certain group of researchers for operating within a closed circle, as mentioned earlier, and then he applauded the Hoffman paper which argued that Raup and Sepkoski's claim of extinction perodicity was invalid. Unusually for a daily newspaper, the *New York Times* has also taken a stance in a scientific debate by expressing hostility to the view of Alvarez and others who favoured an extraterrestrial cause of mass extinctions. The following editorial of 2 April 1985 achieved considerable notoriety, and caused a certain amount of anxiety among impact supporters who held research grants.

Miscasting the dinosaur's horoscope
During the close of the Cretaceous era some 65 million years ago, all dinosaurs disappeared from the earth. Palaeontologists, the students of fossil life forms, have for decades debated inconclusively the reasons for that extinction, but five years ago their game was suddenly snatched away by two brash Berkeley scientists and a crowd of astronomers.

Luis Alvarez, a physicist, and his son Walter, a geologist, contended that a meteorite had slammed into Earth raising such a storm of dust that the sun was blotted out and whole species of animals fell extinct worldwide.

Stretching a provocative idea even further, other scientists claimed to discern a regular pattern in the fossil record: mass extinctions every 26 million years.

The notion of regular extinctions got astronomers excited because the *deus ex machina* required to make giant meteorites crash into earth like clockwork every 26 million years clearly lay in their province. Some posit that an unseen companion of the Sun, christened Nemesis, shakes loose comets each time its orbit passes near a comet cloud. Others contend that the Sun, as it bobs up and down through the plane of the galaxy, is buffeted by comets or dust clouds.

These are rich hypotheses. Why, then, without any further evidence, do they seem so unsatisfying? Perhaps because complex events seldom have simple explanations. Invoking regular squads of meteorites to dispose of the dinosaurs and other vanished species is only to exchange one mystery for another.

On closer scrutiny, the alleged repeating pattern of mass extinctions has faded. The dinosaurs and other vanished species didn't all turn feet up in a day; some were in decline before the end of the Cretaceous. The thin layer of iridium that has been found in many geological strata dating from 65 million years ago could indeed have come from a meteorite, as the Alvarezes suggest, but eruptions of volcanos are now known to be sources of iridium too.

Terrestrial events, like volcanic activity or changes in climate or sea level, are the most immediate possible causes of mass extinctions. Astronomers should leave to astrologers the task of seeking the cause of earthly events in the stars.

An even greater stir was caused by an article on 19 January 1988 entitled 'Debate on dinosaur extinction takes an unusually rancorous turn'. In it the conflict between Luis Alvarez and his opponents was portrayed as being marked by bitter acrimony, with Alvarez uttering a variety of abusive remarks towards his opponents, and scornfully comparing palaeontologists with stamp collectors.[47] This article caused great upset to the Alvarez family, because Luis was dying of cancer at the time and did not know that informal comments made over the telephone were to be made the subject of a piece of sensationalist journalism. In fairness to the journalist, however, who has a reputation for competent and responsible reporting of scientific issues, Alvarez had been making similar remarks for years. During the course of an astonishingly successful career marked by numerous brilliant scientific discoveries and inventions, Alvarez had often exhibited considerable belligerence towards people with whom he disagreed. Though apparently capable of inspiring great loyalty and affection among those who knew him well,[48] tact was evidently not his strong point. Undoubtedly he made enemies among geologists and palaeontologists, who were offended by his arrogance, but he must be held responsible for at least some of the rancour engendered. Where the *New York Times* article is misleading, however, is in its implied suggestion that this rancour has been characteristic of the whole extinctions debate. It would be fairer to say that its tone has been set rather by Luis' more diplomatic son, Walter, who, while arguing his case with vigour, has shown a greater appreciation of the complexities, and maintained good personal relations with his scientific adversaries.

To conclude this chapter I shall attempt to give my personal assessment, for what it is worth, of the current status of the extinctions controversy.

Whether or not extinctions are genuinely periodic, and not many seem to be so persuaded, the idea that they were caused by multiple comet showers receives at the present time little sympathy from the scientific community. The supporting evidence from the stratigraphic record is negligible, and the underlying astronomical theories have been seriously called into question. More plausible earth-bound alternatives exist, of which I personally favour for marine extinctions a combination of regression of epicontinental seas and spreads of anoxic water during the succeeding transgressions.[19] Something very unusual appeared to have happened at the end of the Cretaceous, and opinion is growing that it might have involved an event unique in the Phanerozoic. There is no reason to doubt that major impacts have occurred during the last few hundred million years, but no obvious correlation with mass extinction events exists apart from this. Bolide impact may well have occurred at the K–T boundary, indeed a majority of geologists appears to think so, but the volcanic alternative to accounting for the iridium anomaly and shocked quartz has not yet been conclusively ruled out. Even granted an impact event – or maybe more than one – it is far from clear how it related to the environmental changes that caused extinction. Significant sea-level and perhaps also climatic change and volcanism independent of any impact event seem to be implicated in some way, and it remains a remarkable coincidence that an asteroid should have hit the earth at just this time of considerable environmental disturbance.

The original argument of the Berkeley group implied that alternative processes for the concentration of iridium and related metals, evidence against simultaneity, and evidence contrary to their dust cloud hypothesis would all constitute falsification. In none of these cases, however, have the criteria of evaluation proved clear-cut, or the conclusions universally convincing. Indeed, problems continue to multiply as new data are gathered. In marked contrast to the so-called earth sciences revolution involving the acceptance of plate tectonics, where magnetic anomalies provided the key to the problem and the

consensus of geologists was converted within a few years, over eight
years have elapsed since the Alvarez hypothesis was put forward and
no sign of a consensus about the causes of extinction is yet in sight.
The jury is still out.

Notes

1. Chamberlin, T. C. (1909). *Am. J. Sci.* **17,** 689.
2. Moore, R. C. (1954). *Bull. Mus. Comp. Zool. Harv.* **122,** 259.
3. Newell, N. D. (1967). *Geol. Soc. Am., Special Paper* **89,** 63.
4. Schindewolf, O. H. (1954). *N. Jb. Geol. Palaont. Mh.,* 457.
5. McLaren, D. J. (1970). *J. Palaeont.* **44,** 801.
6. Alvarez, L. W., Alvarez, W., Asaro, F., and Michel, H. V. (1980). *Science* **208,** 1095.
7. Very readable accounts of the development of the hypothesis and its subsequent reception, written in non-technical language, are to be found in the articles by Alvarez *père et fils* (L. W. Alvarez (1987), *Physics Today,* July, 25; W. Alvarez (1986), *Eos* **67,** 649) and the books by Hsü, Muller and Raup (Notes 13, 30, and 9).
8. Alvarez, L. W. *et al., op. cit.* (Note 6).
9. Raup, D. M. (1986). *The Nemesis affair.* Norton, New York.
10. Clemens, W. A., Archibald, J. D. and Hickey, L. J. (1981). *Palaeobiol.* **7,** 293.
11. Smit, J. and Hertogen, J. (1980). *Nature* **285,** 198.
12. Hsü, K. J. (1980). *Nature* **285,** 201.
13. Hsü, K. J. (1986). *The great dying.* Harcourt Brace Jovanovich, San Diego.
14. Emiliani, C., Kraus, E. B. and Shoemaker, E. M. (1981). *Earth Planet. Sci. Lett.* **55,** 317.
15. Ganapathy, R. (1980). *Science* **209,** 921.
16. Orth, C. J. *et al.,* (1981). *Science* **214,** 1341.
17. *Geol. Soc. Am., Spec. Paper* 190 (1982).
18. Bohor, B. F., Foord, E. E., Modreski, P. J. and Triplehorn, D. M. (1984). *Science* **224,** 867.
19. Wolbach, W. S., Lewis, R. S. and Anders, E. (1985). *Science* **230,** 167.
20. Officer, C. B. and Drake, C. L. (1983). *Science* **219,** 1383.
21. Officer, C. B. and Drake, C. L. (1985). *Science* **227,** 1161.
22. Officer, C. B. and Ekdale, A. A. (1986). *Science* **234,** 261.
23. *Geology,* January 1987, 90–92.
24. Officer, C. B., Hallam, A., Drake, C. L. and Devine, J. D. (1987). *Nature* **326,** 143.
25. Alvarez, L. W. (1983). *Proc. Nat. Ac. Sci.* **80,** 627.
26. That this is a misleading and tendentious remark is apparent from a later review by Clemens and his Berkeley colleague, Kevin Padian, who point out that the turnover rate of dinosaurs through their long history

was consistently high. Genera and species went extinct at a rapid rate compared with many other organisms, and the really unusual feature of the end-Cretaceous extinction was that no replacive group emerged.

27. Jastrow, R. *Science Digest*, September 1983, 151.
28. Alvarez, L. W. *et al.*, (1984). *Science* **223,** 1183.
29. Alvarez, W. *et al.*, (1984). *Science* **223,** 1135.
30. Two well-written books outlining the extinction periodicity story, and its possible astronomical implications, are by Raup (*op. cit.*, Note 9) and Richard Muller (*Nemesis – the death star* (1988). Weidenfeld and Nicolson, New York). Raup's book is the more balanced account but Muller's vividly conveys both the excitement that can be generated among groups of scientists following up ideas and the physicist's rather unusual approach to problems of geoscience. The reader should be warned, however, that Muller's enthusiasm for his death star is not widely shared among either geologists or astronomers. This should not be interpreted to mean, though, that he is guilty of *hubris*.
31. Raup, D. M. and Sepkoski, J. J. (1984). *Proc. Nat. Ac. Sci.* **81,** 801.
32. Clube, S. V. M. and Napier, W. M. (1982). *The cosmic serpent*. Faber and Faber, London.
33. Hallam, A. (1984). *Nature* **308,** 68.
34. Bailey, M. E., Wilkinson, D. A. and Wolfendale, A. W. (1987). *Mon. Not. R. astron. Soc.* **227,** 863.
35. Hoffman, A. (1985). *Nature* **315,** 659.
36. Stigler, S. M. and Wagner, M. J. (1987). *Science* **238,** 940.
37. Patterson, C. and Smith, A. B. (1987). *Nature* **330,** 248.
38. A full citation list is provided in Raup and Sepkoski's reply (1988) to Stigler and Wagner's criticisms (*Science* **241,** 94).
39. Sepkoski, J. J. (1987). *Nature* **330,** 251.
40. Hut, P. *et al.*, (1987). *Nature* **329,** 118.
41. Alvarez, L. W. (1987), *op. cit.* (Note 7).
42. Hallam, A. (1987). *Science* **238,** 1237.
43. *Lun. Planet. Inst. Contr.* no. 673 (1988).
44. Indian researchers (A. R. Basu *et al.*, *Eos* **69,** 1487 (1988)) have recently reported the discovery of shocked quartz grains with multiple planar features from a sand layer immediately underlying the lowermost lavas, and argue that this indicates that the Deccan vulcanism resulted from bolide impact. But the age of the oldest Deccan lavas is earlier than the K–T boundary by as much as half a million years, and these results, if confirmed, could be interpreted in a quite different way, that the vulcanism was initiated by an extremely powerful explosive event in the upper mantle.
45. Hoffman, A. and Nitecki, M. (1985). *Geology* **13,** 884.
46. Elizabeth Clemens has made a thoughtful and perceptive analysis of this role in an article in *Social Studies of Science* **16,** 421 (1986).
47. Compare Rutherford's similar remark about all scientists other than physicists (see next chapter, and its Note 5).
48. See for example Muller (*op. cit.*, Note 30).
49. Hallam, A. (1989). *Phil. Trans. Roy. Soc. B*.

8

General considerations

Reflections on people and places

This book is as much about scientists as ideas, or rather about the interaction of scientists with each other and the ideas they have argued about. Indeed, the recurring appearance in the text of leading scientists provides more of a link than the controversies that have been reviewed successively. It is an oversimplification if not a downright distortion to dub these figures radicals or reactionaries with respect to established theory, and any attempt to extend such a one-dimensional classification to those persons involved in more than one controversy is doomed. The transition through an individual lifetime from young turk to old fogey is in some cases clear to see.

Max Planck might have been in an unduly cynical mood when he wrote that 'A new scientific truth does not triumph by convincing its opponents and making them see the light, but rather because its opponents eventually die, and a new generation grows up that is familiar with it'.[1] Nevertheless one is struck again and again by the stubbornness or refusal to recant publicly of leading controversialists, in the face of overwhelmingly adverse evidence.

This is true for instance of leading neptunists like Werner and Jameson. Buch quietly abandoned neptunist views on the origin of igneous rocks without ever acknowledging his debt to his opponents; he never abandoned his catastrophist craters-of-elevation theory. Agassiz, Sedgwick, and Murchison were never converted to uniformitarianism, even less, God forbid, to organic evolution. Kelvin and Joly died without publicly acknowledging that the discovery of radioactive minerals in the crust completely invalidated their geological age estimates, and Jeffreys was as stubborn in resisting the new geophysical evidence indicating lateral movements of continents as he was earlier dismissive of the geological and biological evidence.

Much of this resistance to change is no doubt bound up with pride, with too great an emotional involvement with the ideas which become scientists' babies, and which therefore inhibits the normal functioning of their considerable intellects. It is therefore all the more creditable that Buckland and Lyell were able to publicly abandon their committed positions on major issues. Buckland by successive stages dropped his diluvial theory and became Agassiz' leading supporter in Britain.

Less impressive, Lyell, after decades of opposition to palaeontological progression, conceded the truth of organic evolution. He even made an attempt near the end of his life to come to terms with Kelvin's attack on uniformitarianism.

Porter[2] has stressed the importance of national influence in the development of geological thought in the eighteenth century. This is particularly well illustrated in the succeeding century. Geology in the latter part of the nineteenth century was influenced in Britain by the work of Lyell and the reaction to it to a much greater extent than on the continent. Thus the long debate about the age of the earth, provoked by Kelvin's attack on uniformitarianism, finds virtually no echo in the continental countries.

The modern historiographic approach of assessing the work of scientists in the context of the social and intellectual milieu may, however, be carried too far.

Gould, pondering on why Darwin chose to follow Lyell's doctrine of gradual change so strictly, writes.

On issues so fundamental as a general philosophy of change, science and society usually work hand in hand. The static systems of European monarchies won support from legions of scholars as the embodiment of natural law ... As monarchies fell and as the eighteenth century ended in an age of revolution, scientists began to see change as a normal part of universal order, not as aberrant and exceptional. Scholars then transferred to nature the liberal program of slow and orderly change that they advocated for social transformations in human society. To many scientists, natural cataclysm seemed as threatening as the reign of terror that had taken their great colleague Lavoisier.[3]

Porter asserts that Lyell was committed to 'the brash, rationalist, philosophical conjectural history of the Enlightenment, with its pose of superior detachment, its psychological reductionism, its sardonic contempt for the Middle Ages and for Roman Catholicism'.[4]

While it may be reasonable to maintain that the contemporary *Zeitgeist* cannot be ignored in the development of the human sciences, the connection is far less apparent for the natural sciences. Surely ideas in this sphere may, to some extent, attain their own dynamism, related to an investigation of the natural world provoked by sheer intellectual curiosity, unhampered by socio-political considerations. Thus a gradualistic strand of thought can be traced from Hutton and Playfair via Scrope and Lyell to Darwin, that evolved in opposition to an episodic, catastrophist view of change epitomized by Cuvier and his followers. Marxists have adopted just such an episodic interpretation of historical change involving the so-called dialectical laws, reformulated by Engels from the philosophy of Hegel. Lyell might well have been a country gentleman scared of revolution but is the socially conservative Baron Cuvier now to be regarded as a marxist precursor?

As to be expected from such a youthful nation, serious American involvement in geological thought did not begin until late last century, but has continued to be increasingly important since. As noted in the last chapter the intellectual climate in Germany made it easier for someone like Wegener to develop and promote ideas involving polar wandering and continental displacement than it would have been in the United States, Canada, or Britain. It is all the more intriguing that the modern advances leading to the general acceptance of plate tectonics took place exclusively in these countries; those few Europeans who contributed actively in the early stages had spent a considerable time in the United States. With the last of the great controversies considered in this book, that concerning mass extinctions, the role of American scientists (or scientists working in America) has been overwhelmingly dominant. This has indeed been the case in most other fields of science in the latter third of this century. An interesting sociological point to note is that, in contrast to their relatively conservative attitude towards continental drift, Americans have tended to be more prepared than Europeans to accept, or at least give serious consideration to, the radical idea of catastrophic extinction by bolide impact.

Despite the huge increase in numbers of practising earth scientists over the last two centuries, the history of major geological innovations or controversy can be written without serious oversimplification in terms of a small number of individuals and research centres. What appears to have increased significantly through time is the number of bystanders, for the fact has to be faced that the great majority of geologists and geophysicists are practically-minded specialists lacking an absorbing interest in general concepts. Few, moreover, have had the opportunity or luck to be trained in the technique relevant to a breakthrough, or to be 'in the right place at the right time'.

In Werner's time Freiberg was obviously a geological Mecca, and subsequently Edinburgh and Paris became the leading centres of thought. At the time when the catastrophist–uniformitarian debate was at its height, Oxford and Cambridge universities had become established as major centres of geological research activity after a long period of sloth and indifference to natural science, although the numbers involved were very few. (There was a subsequent relapse, and natural science can only be said to have become firmly established in these august institutions towards the close of the century.)

Subsequently it becomes less easy to identify significant focal points of activity, with the conspicuous exception of Cambridge in the years leading up to the plate tectonics revolution. As with nuclear physics, biochemistry, and neurophysiology in the interwar years, and astronomy and molecular biology after the Second World War, Cambridge was one of the leading centres, if not *the* leading centre, where crucial breakthroughs took place. No other university can rival this record, though one should note that important research bearing on the plate tectonics revolution took place at the University of California at Berkeley, and the impact theory of mass extinctions was propounded here.

One thing that emerges clearly from this general review is the desirability of persistent and persuasive advocacy in order to change thought. It is simply not enough to have a novel idea, however good it may be. The eloquence of Werner and the enthusiasm he transmitted to his star pupils played a major role in establishing neptunism as standard doctrine across Europe and North America. Hutton's work only began to be taken seriously after Playfair translated it into more lucid, concise, and eloquent prose. Lyell was, of course, his own best advocate and a superb one at that. It was not the diffident, scrupulous Charpentier who persuaded the geological world of the correctness of the glacial theory but the more exuberant, forceful, and widely-travelled Agassiz. Kelvin was dauntless in the persistence of his attacks on the uniformitarians.

In the inital stages at least Darwin required Huxley's championing on his behalf, but the eloquence, logical argument, and masterly marshalling of a wide array of supporting evidence in the *Origin* quickly commanded widespread respect and acceptance among other biologists; Wallace's contribution was soon ignored. In a similar way, and presumably for similar reasons, it was Wegener's continental drift hypothesis, not Taylor's, to which attention was paid. Wegener's ideas might have fared better if a capable and enthusiastic advocate such as du Toit had not been, as a South African, effectively isolated from the principal centres of geological research in the northern hemisphere. Certainly the advocacy in the post-war years of those indefatigable travellers, Carey and Runcorn, did a great deal to convert opinion towards continental drift, or at least soften up the opposition, prior to the sea-floor spreading breakthrough. Luis Alvarez was an eloquent and forceful proponent of his impact theory for the end-Cretaceous extinctions and was very adept in handling the media. Indeed, media involvement in the promotion of this idea, and of extraterrestrially-induced periodic extinctions, has been so substantial as to influence the research orientation of a not inconsiderable number of younger scientists.

Lest it be thought that changes in belief are brought about essentially by outstandingly able and persuasive individuals, we must attempt to redress the balance by taking note of the importance of new research techniques.

A major shortcoming in the neptunist-plutonist controversy was the primitive or non-existent stratigraphy. The neptunist's lithostratigraphy was primitive; the plutonists had no stratigraphy at all! The demise of neptunism was brought about as much by development of the powerful new technique of correlating strata by the fossil content as by anything else. The recognition that crystalline metamorphic rocks might be younger in some areas than unmetamorphosed sedimentary strata elsewhere showed its limited validity, to put it mildly.

Similarly, there was a progressive retreat in the catastrophist position in the face of facts unearthed by the application of biostratigraphic principles. What had hitherto been regarded as single paroxysmal events, such as the uplift of mountain ranges, had to be broken up into a series of episodes under the pressure of new data. The success of uniformitarian doctrine owed much to the recognition as a result of the establishment of a trustworthy relative time-scale, that the violence of geological events showed no simple correlation with increasing age. Nor should we ignore the fact that the major weakness of uniformitarianism – its failure to account for biological progression – could only be clearly revealed following the establishment of a sound biostratigraphic scheme.

Secondly, it was the development of uranium–lead dating, based on radioactive decay of isotopes, which transformed our knowledge of geological age and finally undermined Kelvin's arguments after many years of inconclusive debate.

The third outstanding example is the study of rock magnetism, which led in the 1950s and 1960s successively to the plotting of polar wandering paths for different continents and the confirmation of Hess' sea-floor spreading hypothesis. The dramatic transformation of opinion thereby brought about effectively marks the end of the continental drift controversy. Let us take note of the fact that the pioneers in the development of these various research techniques do not, generally speaking, correspond to those who transformed thought as a result of applying them systematically to resolve major geological problems.

Radiometric dating marks the introduction into geology of techniques from the physical sciences, which have become increasingly important through the course of this century. Kelvin may be credited with playing a major role in introducing physical concepts and a distinctive physicist's mode of thinking into geology, though the pioneering efforts of others such as Hopkins should not be ignored.

There has often in the past been some tension between geologists and physicists with a common interest in problems of earth history, which is a natural consequence of their differences in aptitude, training, and outlook. Geologists tend to be staunchly empirical in their approach, to respect careful observation and distrust broad generalization; they are too well aware of Nature's complexity. Those with a physics background tend to be impatient of what they see as an overwhelming preoccupation with trivial detail and lack of interest in devising tests for major theories, and with the geologists' traditional failure to think in numerate terms.

There is indeed a huge gulf between highly numerate scientists like Kelvin and Jeffreys, with their supreme confidence in their own simplifying assumptions and olympian disregard of the inadequately quantitative findings of those who spend their life in detailed study of rocks and fossils, and the traditional survey geologist who is taught to respect above all else what he sees as concrete facts.[5] One of the most gratifying features in the earth sciences after the plate tectonics revolution is the bridges that have been established between geo-physicists on the one hand and geologists trained in the traditional disciplines on the other. There is no longer much sympathy for extremist attitudes, and earth scientists tend nowadays to respect each other's disciplines and cooperate with each other in the pursuit of common goals to a greater extent than ever before.

It is often stated that the plate tectonics revolution came about as a result of advances in geophysics.[6]. This was perhaps most explicitly put forward by Robert Muir Wood in his book *The dark side of the earth* (Allen and Unwin, 1985). He sees geology as practised for well over a century as an out-of-date fuddy-duddy subject characterized by a discredited inductive approach to the gathering of relatively trivial facts, restricted in scope to small areas of the earth's surface. There is little analytical rigour, and the research equipment is simple or even primitive. Not surprisingly its practitioners tend to be conservative and hence distrustful of new ideas. This caricature of traditional geology he contrasts with the modern, scientifically successful approach involving the testing of hypotheses about the earth as a whole, using quantitative methods and sophisticated equipment, and backed up by a thorough grounding in the underlying physics and chemistry.

This thesis may contain a grain of truth, but Muir Wood betrays a fundamental misunderstanding of what has actually taken place. There was always a significant and influential minority of geologists who thought in global terms, or who were sympathetic to the idea of continental mobilism once it had been propounded, while some of the most vehement opposition to continental drift came from an older generation of geophysicists, most notably Harold Jeffreys. In fact it could be argued that the eclectic interests characteristic of the best geologists would make them more susceptible to new ideas than certain mathematical theorists with tunnel-vision. Muir Wood includes stratigraphy and palaeontology within his traditional geology, evidently failing to realize that good stratigraphic correlation by means of fossils is an essential tool in studying the history of the earth as a whole. The sophistication of research equipment is irrelevant. Modern stratigraphic techniques involve the use of geophysical and geochemical as well as palaeontological methods, but the underlying scientific principles of correlation remain the same.

Without the study of fossils the great advances of the justifiably celebrated deep-sea drilling project would not have taken place. Similarly our ideas on geotectonics have been significantly transformed in recent years by the recognition of so-called 'displaced' or 'suspect' .terranes—pieces of land that have travelled extensively in the Pacific before colliding with the Asian or American continental margins. The discovery of displaced terranes came about by research in both palaeontology and rock magnetism. This provides a good example of the point that, in the new earth sciences, traditional and modern research techniques are used in conjunction in the elucidation of the history of our planet. To set these techniques, or their practitioners, into some kind of opposition is both misleading and unhelpful. It is in many ways a pity that we did not follow the advice of Arthur Holmes and subsume geophysics and geochemistry under Geology, understood in its broad sense as the science of the earth. Unfortunately Holmes failed to appreciate the strength of the desire of the new breeds of scientist to establish their place in the sun.

Application to geology of some models of scientific method and change

Most scientists blithely ignore the writings of philosophers and are content to pursue their research interests guided by the traditions established within their subjects. One might wryly add that their work is perhaps none the worse for that.[7] Two writers, however, Karl Popper and Thomas Kuhn, have had an influence extending far beyond the limits of a rather recondite discipline.

One of Popper's most significant achievements in his *The logic of scientific discovery*[8] was to undermine the conventional inductivist view of science dating back to Francis Bacon and which had received its fullest formulation by John Stuart Mill. The process of induction was thought to start with simple, unbiased, innocent observation. Gradually, out of a disorderly array of facts, an orderly theory would somehow emerge. Such an inductivist view was implicitly accepted by geologists.

The essence of Popper's thesis is perhaps most neatly encapsulated in the title of another of his books, *Conjectures and refutations*. We are at liberty to throw up any number of speculative hypotheses but to qualify as scientific theory they must be potentially refutable by means of experiment or observation. Popper emphasized the asymmetry between verification and falsification of scientific theories. In strictly logical terms no amount of confirmatory experiments or observations can be held to verify a given theory, but just one crucial experiment or set of observations can falsify it. Falsification should thus be a primary goal of the good scientist. This view is in sharp contrast to the inductivist position, where the emphasis is upon verification.

Now the word *hypothesis* has in earlier times had a flavour of the wanton or irresponsible about it. If even a great physicist like Isaac Newton could write *Hypotheses non fingo*[9] how much less inclined towards conjecture would be practitioners of the so-called 'descriptive'[10] sciences like geology. An important step towards making hypothesizing more respectable in geology was taken by T. C. Chamberlin, who wrote an influential article near the end of the last century.[11] His 'method of multiple working hypotheses' involved making observations on the natural phenomena under investigation, thinking up several possible hypotheses to account for them, and finally deciding between the hypotheses on the basis of critical observations, perhaps aided by the application of Occam's Razor. The important thing was to approach the natural phenomena initially in a kind of state of innocence, with the mind a *tabula rasa* unguided by a ruling theory.

This is as far as most geologists have traditionally been prepared to go, and Chamberlin's description of how they should proceed in their research has received many an enthusiastic endorsement. The trouble is that nowadays we recognize that observations are theory-loaded and that we need grounds for choice of observations to be made. States of innocence are reserved for the Garden of Eden.

That what Chamberlin called 'ruling theory' is, on the contrary, of paramount importance emerges clearly from consideration of the work of such leading thinkers as Werner, Hutton, Buckland, Lyell, and Wegener. Werner and Hutton, and their contemporaries, reacted so strongly against the speculative theories of the cosmogenists that they overestimated the empirical nature of their own work. Werner went on adding supporting evidence, as he saw it, to a speculative notion dating back to the cosmogenists, allied to a pre-established stratigraphy; in no way does the 'inductive method' make any kind of sense as applied to him. It is also quite evident that Hutton had formulated his revealingly titled *Theory of the earth* long before he discovered some of the most critical supporting evidence, such as angular unconformities or, more particularly, the granite veins cutting schist in Glen Tilt – hence the joy of finding them which puzzled his field guides, but which any creative scientist will understand (see Chapter 1).

Buckland had formulated his diluvial theory before he discovered the best apparently supporting evidence from caves. Agassiz went far beyond the existing evidence when he put forward his glacial theory. Lyell and Wegener continued to add evidence in support of an early-formulated idea over a number of years. It is instructive to note that they were both accused by their critics of being advocates rather than pursuing the proper scientific (i.e. inductive) method.

Nowadays most scientists who think about these things believe that they practise some form of *hypothetico-deductive* rather than inductive method but some consider that Popper might have gone too far. By his rigid demarcation criterion, for instance, theories in the historical sciences, including the theory of evolution by natural selection, would be excluded from true science because falsification in the strictest sense is not possible.[12] A case can indeed be argued that the very foundation of historical geology, the stratigraphic column, is an entirely inductive structure in the Baconian sense.

Although *The logic of scientific discovery* has been widely acknowledged as an intellectual *tour de force*, Popper has been criticized for applying strictly logical criteria too rigorously, and not exhibiting enough knowledge of or insight into the way the operations of science are actually conducted.[13] There are in fact few 'crucial experiments' – they only appear thus in retrospect.[14] In Ziman's opinion[15] Popper's demarcation criterion, that an acceptable scientific theory should in principle be falsifiable, is strategically sound but tactically indefensible. It turns out, in practice, that almost any 'theory' is to some extent 'falsified' by the relevant observations: the question then hinges on whether this failure is to be treated as a genuine objection or whether, pending conceivable improvements in formulation or computation, it may temporarily be overlooked.

Kuhn aligns himself firmly with those, such as Polanyi[16] and Ziman[15] who insist that a proper understanding of science can only be obtained by paying careful attention to how scientists actually operate and interact, which implies considering a whole range of psychological and sociological factors. His basic thesis[17] is that science progresses not by gradual accumulation of knowledge but by radical changes of *paradigm* or world view following longer periods of 'normal science' in which the characteristic activity is the comparatively humdrum one of 'puzzle solving'. Though widely acknowledged as a penetrating and original contribution to the philosophy of science, Kuhn's work has its critics.[18] He has been accused of 'psychologism', that is, paying too much attention to the personal judgements of a consensus of scientists as opposed to purely rational criteria. The charge has been made that he has not given a sufficiently clear definition of what is meant by 'paradigm', and that he has made too sharp a distinction between normal and revolutionary science. Perhaps the most frequently expressed criticism is that Kuhn's monistic view – only one paradigm for a given subject at a particular time – is unduly restrictive.

I find myself in agreement with Ziman[15] when he states that the goal of science is a consensus of rational opinion over the widest possible field, and am content to leave the purely epistemological problems to the philosophers. There is no sure way of achieving certainty or eliminating fallibility, which is far from admitting that scientific beliefs are at the whim of fashion or fancy. Ziman goes to great pains to establish that they are more strictly controlled and rational than any other sphere of thought, by appeals to empirical data, logic, and common sense. Scientific communication is normally more universal and less ambiguous than any other mode of discourse, and is dependent not merely on language and algebraic formulae but on what is termed *pattern recognition*. This refers to diagrams and pictures whose meaning cannot be deduced by formal logical or mathematical manipulation, and is obviously of major importance in sciences such as geology. Scientific knowledge is not so much 'objective' as 'intersubjective'.

How well does Kuhn's model of scientific change fit geology? Some years ago I made a case for the dramatic transformation of thought in the earth sciences community in the 1960s being a good example. There was a comprehensive consensus shift from what I termed a stabilist to a mobilist paradigm.[19] I still think it apposite to refer to this change as a genuine revolution in something like the Kuhnian sense, though we must take note of the fact that both stabilist and mobilist paradigms had coexisted for nearly half a century previously. Hence analogy with political revolution, where transfer of supreme power is total, should not be pressed too far.

With the other controversies discussed in this book, however, the situation is less straightforward. In the case, for instance, of the neptunist–vulcanist–plutonist controversy, it is not easy to define opposing paradigms. As noted in Chapter 1, many vulcanists accepted the overall neptunist stratigraphic scheme and even the idea of a receding ocean; they merely differed on the origin of basalt. It is an oversimplification to distinguish the neptunists and plutonists respectively as supporters of the role of water and internal heat as the dominant geological agent. Water was as important as heat in the Huttonian system. Both paradigms were incomplete as world views in this primitive stage in the science, most notably because of the deficiency or total absence of stratigraphic framework. The demise of neptunism, which took place with a whimper rather than a bang, was due at least as much to the development of a satisfactory means of stratigraphic correlation as to the triumph of plutonism.

What then of the catastrophist–uniformitarian controversy? Here is surely a clear difference in world view. However, we have to be more precise, distinguishing in the first place the common ground of Lyell's methodological uniformitarianism from his more contentious substantive uniformitarianism, and in the second place the directionalist theme in catastrophist theory from the episodicity theme, whereby long periods of stability are periodically interrupted by episodes of comparatively abrupt change. Whereas the cruder directionalist view associated with Lyell's critics, from Conybeare to Kelvin, was ultimately discredited and Lyell substantially vindicated by early this century, the change in thought was anything but revolutionary. To pursue the political analogy, it was more a matter of slow reform, with temporary setbacks such as Kelvin's earlier attacks. Furthermore Lyell was shown to be wrong on the question of organic progression. As for the episodicity theme, we have seen a revival in recent years, as noted in Chapter 5, albeit in a more sophisticated form than put forward in the early nineteenth century.

It might be thought that the Ice Age controversy is not of sufficiently general significance to qualify in a discussion on the application of Kuhnian theory. After all it could be, and was, embraced by catastrophists and uniformitarians alike. However, the concepts of marine drift and land ice are general ones, from which can be generated more detailed hypotheses by which to account for particular phenomena observed in the field. If, however, we talk of *drift* and *land ice* paradigms we must accept that they coexisted for several decades before the land ice paradigm became almost universally accepted. Lyell, for instance, could quickly welcome Agassiz' demonstration of the ubiquity of land ice phenomena in his native

Scotland without feeling the immediate need to abandon the idea of widespread inundation and drifting icebergs which melted to dump their sedimentary load. Once more the change in consensus view was a case of gradual reform rather than revolution.[20]

Illuminating and thought-provoking though Kuhn's work is, I prefer the account of scientific change outlined by Imre Lakatos,[21] a philosopher who deserves to be more widely read by scientists.

Lakatos argues that scientists are guided by what he terms *research programmes*, which consist of methodological rules, some indicating which paths of research to avoid (*negative heuristic*) and other which paths to pursue (*positive heuristic*). All scientific research programmes may be characterized by their 'hard core'. The negative heuristic forbids us to attack this hard core, protected by a belt of auxiliary hypotheses, which must bear the brunt of critical tests and which must be adjusted or replaced if necessary to defend the core of beliefs. A research programme is successful if all this leads to a 'progressive problem shift', unsuccessful if it leads to a 'degenerating problem shift'.

Newton's gravitational theory is cited as the classic example of a successful research programme. In the Newtonian programme the negative heuristic protects the hard core of the three laws of dynamics and the law of gravitation, which are irrefutable by the methodological decision of its protagonists. In the early years numerous apparent anomalies and counter instances were successfully converted into corroborating instances and the threat of 'defeat' turned into 'victory'; meanwhile the empirical content increased progressively.

The positive heuristic of the programme, which dictates the long-term research policy, consists of a partly articulated set of suggestions or hints on how to modify and render more sophisticated the 'refutable' protective belt, and may lay out a programme that lists a sequence of ever more complicated models simulating reality; it can forge ahead with almost complete disregard of 'refutations'.

It seems to me that Lakatos' account, of which the foregoing is a very inadequate digest, gives a more penetrating description of scientific activity than either Popper or Kuhn. Most scientists are not engaged for the bulk of their professional lives in either dreaming up or refuting radical new theories. On the other hand, Kuhn's term 'puzzle solving', for the dominant periods of 'normal science', appears to trivialize their activity. There may indeed be times when one paradigm quickly gains ascendency over another, as in the earth sciences 'revolution' in the 1960s, but Lakatos allows more than one paradigm to coexist. One may well be in a progressive, another in a degenerating phase, as new observations or experiments tend to corroborate one set of auxiliary hypotheses and refute the other, or two coexisting paradigms may have an equal claim to serious attention. For such cases the term 'revolution' is clearly inapposite. Likewise, Chamberlin's method of multiple working hypotheses appears sensible enough at a lower level of theorizing. The underlying theory, corresponding to Kuhn's paradigm, operates at a higher, or at least more general, level and is less amenable to direct challenge.

One can criticize aspects of the Lakatos model. The choice of the term 'research programme' is a little unfortunate because of the much more specific connotation it has for most working scientists. Is the hard core of belief really so inviolate from direct challenge and can it not change with time under pressure from criticism? Nevertheless I am attracted by its emphasis on the hierarchial character of scientific theorizing, and it throws new light on the great geological controversies.[22]

Thus there was no critical, falsifying set of observations that undermined neptunism. Werner accommodated newly-discovered facts by introducing auxiliary hypotheses, such as episodes of temporary return of the universal ocean. Neither d'Aubuisson nor Buch, on recognizing that the Auvergne basalts were igneous in origin, were provoked into immediately extrapolating their results to the more crucial basalts of Saxony, even less to abandoning the overall neptunist interpretation of geological history. It was not necessarily stubbornness or stupidity which prompted Richardson to reject the suggestion that the detailed observations on Salisbury Crags had any critical bearing on a major theory of the earth.

As time passed, of course, more and more anomalies appeared that could only be fitted into the overall theory by increasingly implausible *ad hoc* hypotheses, such that there was a gradual waning of belief. In Lakatos' terminology, neptunism in the early years of the nineteenth century had entered the phase of degenerating problem shift.

Something not dissimilar happened in the case of the stage-by-stage retreat of the catastrophist school of thought. Diluvialism was abandoned early, and progressively less emphasis was placed on events of paroxysmal intensity which could not be matched at the present day. Eventually even the directionalist position was undermined, at least for the Phanerozoic. In the case of the continental drift controversy geologists antagonistic to Wegener attempted to answer one of his most serious challenges by doing away with vast, sunken continents beneath the oceans in favour of much more restricted 'isthmian links'. This particular auxiliary hypothesis, protective of the hard core of stabilist belief but accommodating of the biogeographic evidence of former intercontinental connections, could only decisively be ruled out by oceanographic research after the Second World War. A further example concerns the mass extinctions controversy. As supporters of the Alvarez impact hypothesis came to recognize that a single impact event could not account for many of the end-Cretaceous extinctions, which must have been spread over an appreciable period of time, a modified and less falsifiable theory of multiple comet impact was put forward.

The future of plate tectonics

I assume that few will dissent from the view that, by normal scientific criteria, plate tectonics has proved a highly successful theory. It has wide scope, considerable explanatory and predictive power and a beautiful first-order simplicity. It accounts for the wealth of ocean-ographic data brought to light since the Second World War in a way that the deposed paradigm denying continental drift conspicuously failed to do. Hypotheses based on the theory have survived a diversity of tests involving, for instance, seismology, deep-sea drilling, and the divergence through time of different continental polar wandering paths. Apparent anomalies, such as the uncrumpled young sediments in certain deep-sea trenches, have been satisfactorily explained away.

Are we now the victims of this success—can the younger generation of earth scientists anticipate future breakthroughs of comparable magnitude? It is certainly true that what one might call the 'surprise factor' has diminished in the literature of recent years. More and more researchers are treading quite well-worn paths and having to work harder and in more detail to discover something of real novelty. In fact many research fields seem to have achieved the stage of 'normal science', if we are to continue to use Kuhn's term. For instance, a research student at a major oceanographic institution might be given the task of analysing in detail a small sector of ocean floor, which could involve sorting out some little local problems that emerged from an earlier reconnaissance study of magnetic anomalies, acting on the assumption that plate tectonics is true. This is obviously puzzle-solving in the Kuhnian sense.

If there were nothing more to it than that one could hardly expect the imagination of a wider public to be deeply stirred for long, and research funding sources might become increasingly reluctant to pay an ever more expensive bill. But of course there is much more to it. When one considers the whole range of geology, one can hardly fail to be impressed at the way in which plate tectonics has thrown new light on, for instance, mountain-building, igneous and metamorphic processes, the formation of metallic ore deposits, eustatic changes of sea level, palaeoclimatology and palaeobiogeography. These and other research fields have been revitalized by the new paradigm and I believe that a good case can be made for arguing that not just the quantity but also the quality of the relevant research has increased. In other words we tend nowadays to operate at a higher level of understanding than in the old 'stabilist' days, which is not to deny the numerous problems and uncertainties that lie ahead, or the marked differences in quality of particular pieces of research.

Now this impressive progress in geology has depended on the implicit assumption that plate tectonics is true. Utilizing the Lakatos terminology, one could say that the plate tectonics research programme is in a progressive phase, with the hard core of basic belief being protected from direct assault by a series of auxiliary hypotheses that are still being generated. Some of these are being corroborated by new data, others refuted, but the fundamental tenets have of necessity remained immune.

Such a state of affairs should be very worrying to a strict Popperian, because it appears to imply that plate tectonics is not directly falsifiable and hence not scientific. The limitations of this stark approach have been outlined above and I would rather use Ziman's more modest consensus principle in conjunction with the Lakatos model. The dramatic switch of a consensus of earth scientists to a mobilist paradigm is now a fact of history. What we are entitled to ask now is: can we conceive of a further consensus switch in the future, away from plate tectonics? Is it just another fallible theory that will have to be abandoned as knowledge increases, or will it serve as the unshakeable core of earth science for the foreseeable future?

It should be evident that, if plate tectonics is to be rejected in the future, a whole series of awkward new discoveries in a range of disciplines must be made, such that the protective belt of auxiliary hypotheses cannot, in the collective judgement of the geological consensus, cope with the anomalies that arise. Forecasting the future is notoriously hazardous, but I am willing to speculate for the purpose of provoking discussion.

From the foregoing comments it should be evident that I consider it highly unlikely that there will be any dramatic change in the immediate future. Even if anomalies emerge and accumulate, plate tectonics has been far too successful in recent years to be abandoned without intensive efforts being made to explain them away within the framework of existing theory. For the longer term let us consider three possibilities.

1. Despite the heroic efforts of A. A. Meyerhoff and a few others I see no prospect whatever of a counter-revolution to a stabilist paradigm denying continental drift. Many of the objections raised to plate tectonics have proved unconvincing to the geologic consensus; others can be readily accommodated by slight modifications to current theory.

2. It is conceivable, though unlikely, that plate tectonics could be replaced by another mobilist paradigm, that invoking an earth that has expanded rapidly since the early Mesozoic.[23] This paradigm, which can perhaps account for many of the new oceanographic data as well as plate tectonics, differs fundamentally in denying the basic tenet that the amount of crust created at oceanic ridges equals the amount destroyed in subduction zones. One could imagine a research programme devoted to demonstrating, for instance, that extrapolation back through time from the existing pattern of ocean floor magnetic anomalies leads to a conclusion inconsistent with the preservation of constant earth radius; also that the phenomena of active continental margins can be satisfactorily explained without recourse to subduction, and that there must have been continuous continental crust across Tethys, an ocean gap accepted today by the majority of geologists.

There are, however, serious geophysical objections to overcome, and other major geological phenomena satisfactorily accounted for by plate tectonics must be plausibly explained away. These include orogenic and igneous activity prior to the Mesozoic, and the huge late Cretaceous marine transgression. At present the model of rapid earth expansion in the recent geological past appears to create more problems than it solves, and some would consider that it can now be dismissed out-of-hand.[24]

3. By far the likeliest possibility is that plate tectonics will continue to be accepted by the consensus, but perhaps with some significant modifications. Thus Molnar[25] has pointed out that its basic tenet, the rigid-body movements of large plates of lithosphere, fails to apply to continental interiors, where buoyant continental crust can detach from the underlying mantle to form mountain ranges and broad zones of diffuse tectonic activity. A further example is provided by the work of Jordan.[26]

Jordan points out that while existing theory had been very successful in describing the production and destruction of basaltic ocean crust, seismological data reveal substantial contrasts between subcontinental and suboceanic mantle extending down for several hundred kilometres, implying deep root zones that travel coherently with the continents. This challenges a basic tenet of plate tectonics, that the plates are confined to the lithosphere, which averages a mere 100 km in thickness, and slides over a weaker asthenosphere signifed by the low velocity zone.

Jordan goes on to develop a speculative model, consistent with seismological and petrological data, involving basalt depletion of the upper mantle beneath the geologically ancient shield areas or cratons. While not able to judge the technical merits of this model, my non-specialist reaction is that it is plausible and ranks as just the kind of constructive modification of existing theory that one can reasonably anticipate in the future – the apple cart is not thereby upset. Ironically, somewhat similar arguments based on seismic and heat flow data were put forward in pre-plate tectonics days by Macdonald,[27] who inferred that the upper mantle was significantly different under the continents and oceans. Macdonald's conclusion that this denied the possibility of continental drift was nevertheless rejected and his data later reinterpreted by others who accepted plate tectonics. Such is the power exerted by the dominant paradigm.

We are probably only at the beginning of a long debate about the character of the mantle, improved knowledge of which is obviously vital to a proper understanding of the forces controlling plate motions. Without this plate theory is bound to remain incomplete. Failure to provide a plausible mechanism was one of the major reasons why Wegener failed to convert the consensus to his mobilist views. Are we that much the wiser today?

Notes

1. Planck, M. (1949). *Scientific autobiography and other papers*, p. 33. Translated by G. Gaynot. Phil. Lib., New York.
2. Porter, R. (1977). *The making of geology*. Cambridge University Press.
3. Gould, S. J. (1980). *The Panda's thumb*, p. 180. Norton, New York.
4. Porter, R. (1976). *Brit. J. Hist. Sci.* **9**, 91. In his determination to discredit Lyell's historical account in his *Principles* Porter gets quite carried away. 'Because of his didactic polemical intentions, Lyell was to some extent writing anachronistic Whiggish heroes-and-villains history at its most opportunistic. Lyell is too preoccupied with praising (faintly) the good and satirising the puerile to have any consistent concern with investigating why men thought the way they did, or why some arrived at the right answers while others "retarded the progress of truth" – at least such explanations as he offers are glib caricatures.'

    I do not think that this is a fair assessment. Lyell is generous to many of his predecessors and, in the light of contemporary knowledge and attitudes, reasonably balanced in his judgements. His attack on the neptunists is expressed in more temperate language than that of many contemporaries and, indeed, of Porter.

5. Rutherford may have had his tongue in his cheek when he uttered his notorious dictum, 'All science is either physics or stamp collecting', but Kelvin would doubtless have earnestly approved of this reductionist attitude, for he is on record as saying that data that cannot be quantified are hardly worthy of a scientist's attention. Fortunately this traditional physicists' arrogance towards other sciences has declined in recent years, though Luis Alvarez can perhaps be considered a throwback.

6. While this may be true of the key techniques, we should bear in mind that Hess and Vine took their undergraduate degrees in geology, while Tuzo Wilson, though he was for many years Professor of Geophysics at Toronto, has always written like a good old fashioned tectonic geologist – one never sees an algebraic equation in his papers!

7. That, however, scientists' beliefs may be strongly influenced by their philosophy of science is persuasively argued by Michael Ruse (1979, *The Darwinian revolution*, Univ. Chicago Press) with respect to the contrasted attitudes of Herschel and Whewell towards Lyell's doctrine. John Herschel, one of the most distinguished astronomers, and indeed all round scientists, of his day, enthusiastically welcomed the *Principles* because Lyell's thought was in line with his own doctrine of *verae causae* (true causes), demanding argument analogically from our own experience. This support was decidedly ironic because Herschel's own research on the change of orbital eccentricity through time supported the directionalist notion of a cooling earth. Whewell, on the other hand, showed himself to be a rationalist rather than an empiricist. He felt that the *vera causa* principle unduly limits any methodological rule based on it, because 'it forbids us to look for a cause, except among the causes with which we are already familiar. But if we follow this rule, how shall we ever become acquainted with any new cause?'

8. Popper, K.R. (1959). *The logic of scientific discovery*. Hutchinson, London.

9. 'I do not think up hypotheses.'

10. As opposed to 'analytical', of course, but the term 'descriptive' applied to geology makes me wince, as though interpretation were not an integral part of the science.

11. Chamberlin, T.C. (1890), *Science* **15**, 92.

12. Popper has recently backtracked on this position, and now accepts that historical theories can be considered scientific as long as they are testable. (Popper, K.R. letter in *New Scientist*, 21 August 1980).

13. For articles critical of Popper see Suppe, F. (ed.) (1977). *The structure of scientific theories* (2nd. edn), University of Illinois Press, Urbana.

14. Lakatos, I. (1978). Anomalies versus 'crucial experiments'. In *Mathematics, science and epistemology: philosophical papers of Imre Lakatos* (ed. J. Worrall and G. Currie). Cambridge University press.

15. Ziman, J. (1978). *Reliable knowledge*. Cambridge University Press.

16. Polanyi, M. (1958). *Personal knowledge.* Routledge and Kegan Paul, London.
17. Kuhn, T.S. (1962). *The structure of scientific revolutions.* University of Chicago Press.
18. See, for instance, Lakatos, I. and Musgrave, A. (eds.) (1970). *Criticism and the growth of knowledge.* Cambridge University Press.
19. Hallam, A. (1973). *A revolution in the earth sciences.* Oxford University Press.
20. A non-revolutionary pattern of change for 18th and early 19th century geology is also perceived by Rachel Laudan (*From mineralogy to geology: the foundation of a science, 1650–1830*, Chicago University Press, 1987), who sees progress in geology through this time as being controlled by 'guiding assumptions' rather than by over-arching theory. Similarly Greene (*Geology in the nineteenth century*, Cornell Univ. Press, 1982) infers that 19th century research in geotectonics advanced without there ever having been an adequate comprehensive theory, and recognizes neither linear progress nor revolutions.
21. Lakatos, I (1970). Falsification and the methodology of scientific research programmes. In *Criticism and the growth of knowledge* (eds. I. Lakatos and A. Musgrave) p. 91. Cambridge University Press.
22. Secord (*Controversy in Victorian geology: the Cambrian–Silurian dispute*, Princeton University Press, 1986) goes so far as to reject totally any attempt to apply models such as Kuhn's or Lakatos' to the Cambrian–Silurian controversy, arguing that it is more desirable to investigate in fine detail in the tradition of the modern historian the social as well as scientific interactions of individual scientists and their affiliations and allegiances to different types of institution. While such an approach uncovers much illuminating detail, there is a danger of missing the wood for the trees. Secord's example is too limited in scope and involvement in theory to qualify for refutation of general models of scientific change. One must always bear in mind that the more general a model, the more likely it is to be oversimplified as a representation of phenomena, but it may acquire a correspondingly greater heuristic value.

    Le Grand, in his book *Drifting continents and shifting theories* (Cambridge University Press 1988) believes that general models of scientific change can be applied with success to the history of the continental drift controversy, but prefers the model of Larry Laudan to those of Kuhn and Lakatos. Laudan (*Progress and its problems: towards a theory of scientific growth*, University of California Press 1977) argues that the fundamental aim of science is to maximize the scope and number of solved empirical and conceptual problems while minimizing the number of anomalies and conceptual problems generated. His *research traditions* evolve over time, and the theories that express them also change over time. There is competition between research traditions and, within them, competition between theories. Scientists may have a strong preference for one or the other but are not trapped within a single paradigm. I am now inclined to share Le Grand's view that Laudan's model approximates closer to reality than those of his rivals.

23. Carey, S. W. (1976). *The expanding earth*. Elsevier, Amsterdam, Owen, H. G. (1976). *Phil. Trans. R. Soc. London* **A281**, 223.
24. A. Hallam (1984). *Geol. Mag.* **121**, 653. R. Weijermars (1986). *Phys. Earth Planet. Int.* **43**, 67.
25. P. Molnar (1988) *Nature* **335**, 131.
26. Jordan, T. H. (1979). *Sci. Am.* **240**(1), 70.
27. Macdonald, G. J. F. (1963). *Rev. Geophys.* **1**, 587.

Index